Kister Distillation Symposium 2015 (Abstracts)

Topical Conference at the 2015 AIChE Spring Meeting and 11th Global Congress on Process Safety

Austin, Texas, USA
26-30 April 2015

ISBN: 978-1-5108-0696-2

Printed from e-media with permission by:

Curran Associates, Inc.
57 Morehouse Lane
Red Hook, NY 12571

CURRAN ASSOCIATES INC.
proceedings.com

Some format issues inherent in the e-media version may also appear in this print version.

Printed by Curran Associates, Inc. (2015)

AIChE
3 Park Avenue
New York, NY 10016-5991

Phone: (203) 702-7660
Fax: (203) 775-5177

www.aiche.org

Additional copies of this publication are available from:

Curran Associates, Inc.
57 Morehouse Lane
Red Hook, NY 12571 USA
Phone: 845-758-0400
Fax: 845-758-2634
Email: curran@proceedings.com
Web: www.proceedings.com

TABLE OF CONTENTS

409505 Chemical Engineering: Is This the Golden Age?

Abstract:

Schlanger will offer his insights on the current opportunities in the chemical process industries in his talk entitled “Chemical Engineering: Is This the Golden Age?” on Monday morning, April 27. At the conclusion of his lecture, Schlanger will receive the AGILE Award in recognition of his leadership at CEVA Group, a $9 billion logistics supplier to industry worldwide; LyondellBasell, a major player in the plastics, chemicals, and refining industries; Arco Chemical; and at the other companies where he has held leadership positions.

402146 Bibliography of Dr. Elzear Lemieux

Abstract:

Dr. Elzear Lemieux had a B.S. degree in Chemistry from Assumption College and M.S. and Ph.D. degrees in Chemical Engineering from Worcester Polytechnic Institute.

Dr. Lemieux began his career at M.W. Kellogg in 1950. Throughout his career, he was primarily involved the design of trays and packings, and vessel design, in general. He was a strong proponent for the use of sieve trays in the chemical processing industry.

He and his team performed extensive testing of sieve tray capacity at the M.W. Kellogg laboratory in Piscataway, New Jersey. These tests studied and quantified the effects of hole diameter, tray spacing, weir rate, and vapor rate on tray capacity and they also led to an organized set of correlations for sieve tray capacity and downcomer backup.

Dr. Lemieux was a supporter of FRI since its inception. He served on the FRI Technical Advisory Committee (TAC) from 1960 through 1975 for the M.W. Kellogg Company. He was extremely interested in basic fundamental research on tray and packing capacities and efficiencies. He retired from M.W. Kellogg in 1982, but continued to support FRI and attended the FRI TAC meetings. He was Kellogg's screener of FRI Topical and Progress Reports, until he stopped consulting for Kellogg in 1991.

He and his wife of 65 years, Claire, were very active members in the community and especially in their church, St. John Vianney Catholic Church in Houston. His wife passed away in February 2013.

Dr. Lemieux passed away in November 2013.

This presentation describes some of Dr. Lemieux's contributions to separations engineering.

400263 Flow-Induced Vibration of Trays

Abstract:

Abstract for 2015 Spring AIChE meeting

Flow-induced vibration of trays

RC Darton, W Zhang and PH Taylor

The paper presents a mechanism that explains the vibration of distillation and absorption trays at low throughput that can lead to mechanical failure. As vapour flows through the perforations in the tray, a Helmholtz resonance occurs in which oscillating liquid near to each perforation cyclically exchanges kinetic energy with potential energy (head) in the bulk liquid across the tray. The rate of exchange is determined by the wave speed in the bulk liquid. This leads to an oscillation in pressure drop which can resonate with the mechanical structure, leading eventually to fatigue cracking and failure. We make recommendations based on this analysis, which will help to prevent such vibrations. Tray damage, which can occur with both valve and sieve trays, can be both hazardous (cracked column shells can result) and expensive. Our recommendations relate both to design and operation of columns.

402148 Influence of Tray Geometry and Hydraulic Conditions on Vibration Induced Damages of Sieve and Valve Trays

Abstract:

The present paper is a summary of observations on failures of operating trays in various industrial applications. The affected trays include fixed valves (type SVG and SRV), movable valve trays (V1-valves and A3 cage valves) as well as sieve trays.

The different incidents are investigated to derive general indications that promote vibrations and – as a consequence - failures of the trays. The failures are analyzed with regard to valve types, special design features as well as special operating conditions that lead to the failure. The basis are operating data and pictures of the damaged trays.

Also the failure modes of the trays are investigated in detail to derive the destruction mechanism. Where possible video sequences through a sight glass were taken in operating columns to visualize gas and liquid flow as well as vibrations. These findings were compared to experimental results in a pilot plant with a transparent column.

General design features that promote vibrations in combination with special operating conditions like low liquid load could be identified. The position of all operating points in the operating range has to be carefully investigated with respect to parameters that influence tray vibrations.

Based on these findings, guidelines for potentially critical trays were derived. These include mechanical design criteria as well as process and hydraulic design guidelines.

402149 FRI Test Results and Industrial Applications for New High Capacity Raschig Mini Fixed Valve

Abstract:

Raschig Mini Fixed Valve trays have been applied successfully in various industrial applications. The most recent finished FRI-tests support the superior capacity of this valve compared to other fixed valve tray designs as well as sieve trays. During the presentation the test results will be provided, as well as grass-root design and revamp cases will be discussed.

402150 Troubleshooting and Solving a Gas Treater Downcomer Unsealing Problem

Abstract:

ADGAS have commissioned the Integrated Gas Development (IGD) project that transfers 900 MMSCFD of dry gas from the gas wells to the Habshan Industrial site through a 30" pipeline. The project comprises a gas receiving station, three gas drying units (300 MMSCFD each), three GT driven compressors, and a fuel gas treatment (FGT) system with the capacity of 25 MMSCFD of sweet gas. The FGT system was designed by an engineering contractor from Japan and the tower internals by a major tray vendor.

Since the start-up/commissioning of the FGT system it was observed that when the plant operates above 10 MMSCFD, the absorber column became very unstable. A sudden increase in pressure drop in the Fuel Gas Treater column occurred, and that resulted in a sudden level rise in FG KO drum, tripping the whole system. Changes to the MDEA flow rates did not appear to have much influence. An attempt to fix the problem by the contractor and vendor failed to provide a significant improvement.

A thorough troubleshooting investigation by ADGAS, together with hydraulic analysis by Fluor, yielded a correct diagnosis and a successful solution.

Downcomer seal loss in moving valve trays was the root cause of the flooding and instability problem. The constant dry pressure drop held by the valve tray as valves close off at reduced rates can make downcomer sealing difficult in moving valve trays, especially with large clearances. Large downcomer clearances should be avoided at low liquid rates, both with valve trays and other trays.

The valve trays in the tower were replaced by the new bubble cap trays. The tower has been performing well ever since with no signs of flooding or instability, even well above the design rates.

399966 Optimizing Preflash Option for Light Tight Oil Processing

Abstract:

The basic function of crude distillation units is to provide the initial separation of crude oil feed into the desired fractions to feed to downstream units. Newly introduced light tight feed processing may limit existing crude distillation unit capacity. Addition of feed pre-flashing option may be one of feasible options to improve crude distillation unit capacity for light tight oil processing.

Two preflash options which are commonly implemented to crude distillation unit are (1) preflash drum and (2) preflash columns. Each option has strengths and weakness.

This paper introduces improved preflash system which optimizes traditional preflash drum and preflash column option. The refinery economics are maximized through optimum OPEX and CAPEX. The case study including revamp economics will be demonstrated.

CH2MHILL.

Atmospheric Preflash Towers

Andrew W. Sloley – CH2M HILL, Bellingham, WA

Prepared for Presentation at the AIChE 2015 Spring Meeting, Austin, TX, 26-30 April, 2015

An Introduction to Preflash Towers

2

Preflash Drums and Towers

- Equipment upstream of the atmospheric distillation tower
- Allows vaporization of lighter material
- Most often added to existing units for capacity

CH2MHILL.

Unit Survey – 50 Units

Configuration	Units
No preflash	23
Preflash drum	11
Preflash tower	16

Tower Config	16 Units
Steam Stripped	4
Reboiled	5
Flash Drum	10

Tower Type	16 Units
Independent	11
Shared overhead	2
Complex	3

Tower Heat	16 Units
Feed – Exch only	14
Feed – Fired	2
Reboiled	5
Steam Stripped	4
Fd Fired + Reboiled	1

CH2MHILL.

Basis

5

Basis Chosen

Crude	°API Gravity	Specific Gravity
BCF 24	23.3	0.9141
Tia Juana Lt	31.6	0.8676
Bakken	42.1	0.8151
Gippsland	49.7	0.7810

- Feed enthalpy
  - Constant heater outlet temp
- Other
  - Heater outlet vaporization
- Atmospheric Overhead
  - Temperature constraints
  - Reduces corrosion and makes overhead heat integration practical
  - ~70% of units have heat recovery from overhead
- Atmospheric Products and Heat Recovery
  - Economic optimization
- Lowest side-draw
  - AGO or diesel as lowest product
- Equipment limits
  - Internal liquid rates

CH2MHILL. 6

Configuration Options

7

Configuration Options and Possible Operations

- Steam Stripping
  - Preheat feed to flash
  - Recover light material from bottoms by vaporization with reduced partial pressure
- Reboiled
  - Preheat feed to flash
  - Recover light material from bottoms by fractionation
- Preflash
  - Preheat feed to flash
- Operations
  - Light naphtha overhead
  - Full range naphtha overhead – used in comparison
  - Others possible

CH2MHILL.

8

Base Unit (Topping Unit)

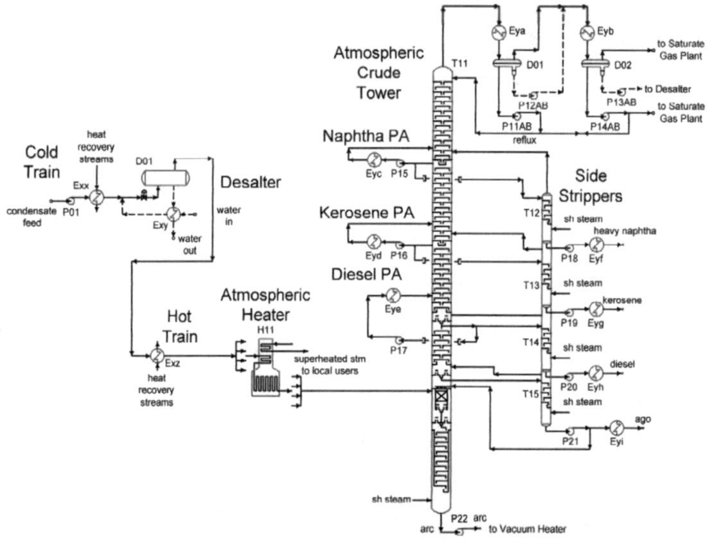

Steam Stripping

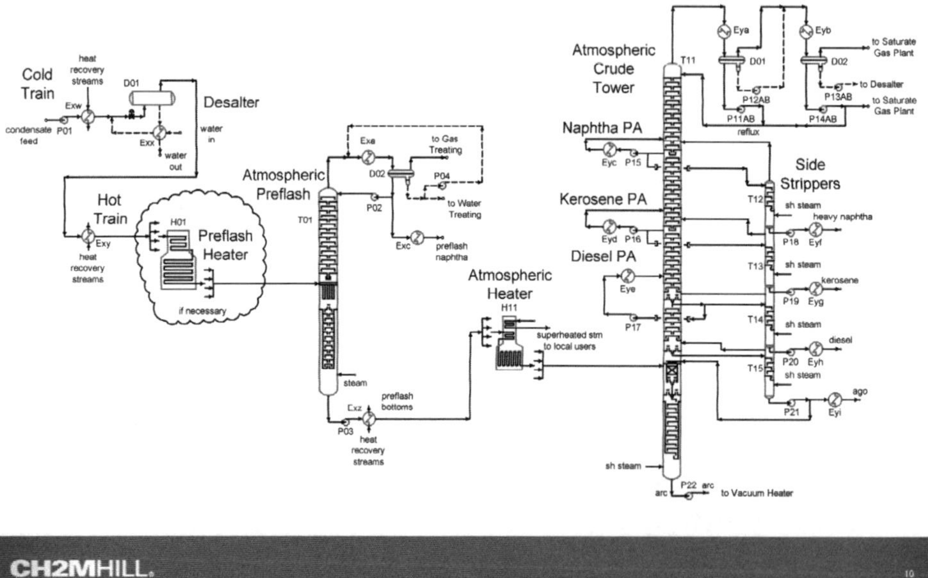

Reboiled

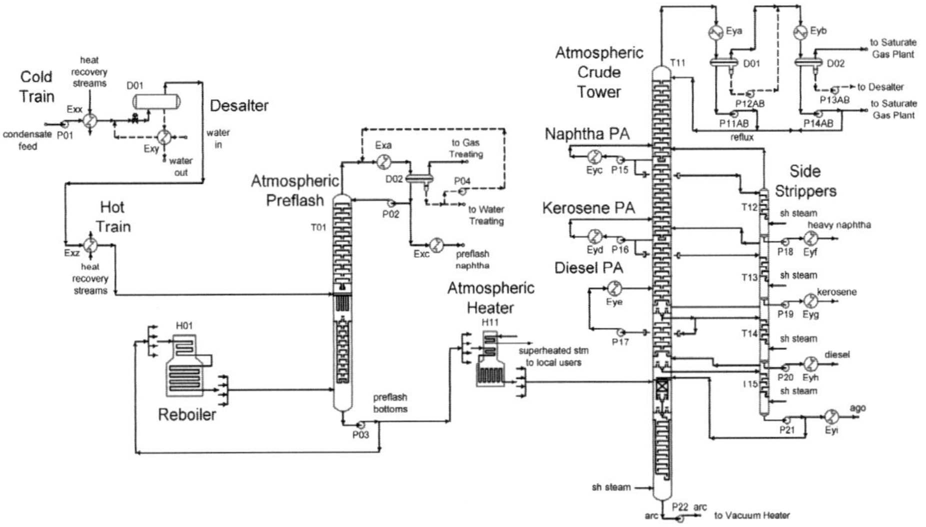

Flash Zone

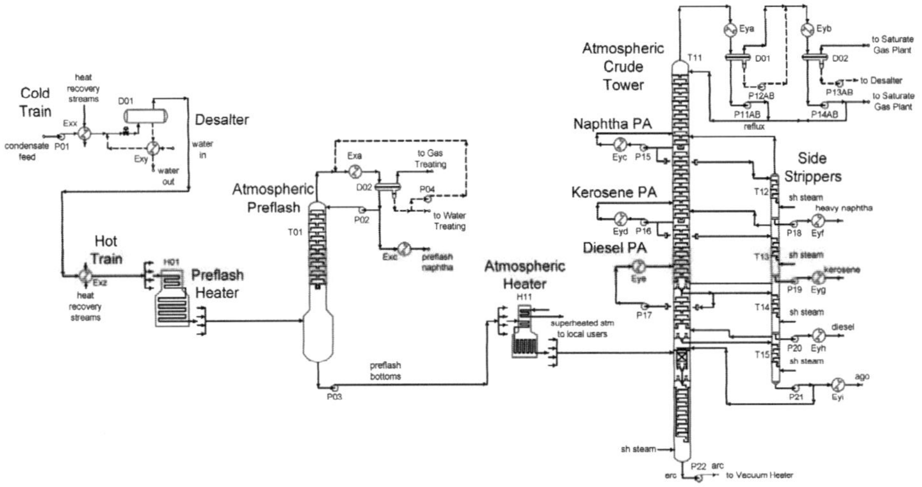

Comparisons

Comparisons Made

- Four crudes
- Four configurations – topping plus three preflash
- Economic optimum for heat recovery and yields
  - Re-optimized for each case
  - Conditions for maximum distillate (kerosene plus diesel) recovery
- 16 total units evaluated

Evaluation Basis – Key Points

- Added to an integrated atmospheric-vacuum unit
- Heat integration and desalter performance vary with crude
- Preflash naphtha cutpoint kept constant
- Equipment constraints for good practice imposed on
  - Minimum exchanger duty
  - Fractionation liquid and vapor rates
  - Fractionation temperature profiles to avoid corrosion
  - Fired heat outlet temperatures and feed vaporization

CH2MHILL. 15

Fractionation Stages versus Reflux

- Preliminary work showed very strong driver towards maximum trays
  - Maximum trays = minimum reflux
  - Minimum reflux = minimum energy downgrade
  - True even with low energy price level of $4 per Mbtu gas price (3.79 $ GJ)
  - More expensive energy makes this driver even more powerful
- 15 theoretical stages used in all rectification sections
- Sensitivity analysis showed 15 stages best overall choice
- Same number of stages used to generate TBP data via D2892 method

CH2MHILL. 16

Stripping Trays versus Reboiled Trays

- Same physical number of trays used in the steam stripping configuration versus reboiled configuration
- This approach gives an approximate equal investment analysis
- Steam stripping trays have much lower efficiency
- Result is
  - Steam stripping sections – 3 theoretical stages
  - Reboiled stripping sections – 9 theoretical stages

CH2MHILL.

17

Results and Conclusions

18

Constraints

- Most constrained by equipment and process limits
- 12 preflash cases examined
  - 9 had some optimization possible
  - 7 could vary kerosene pumparound duty
  - 2 could vary heavy naphtha duty (Bakken)
  - 3 were at constraint limits and required variation of feed temperature with a fired heater (flash zone)

Constraint Summary

Feed	BCF24	BCF24	BCF24	BCF24	TJL	TJL	TJL	TJL
Configuration	Topping	Stripped	Reboiled	Flashed	Topping	Stripped	Reboiled	Flashed
No Preflash	X				X			
Overhead duty set by overhead temp	X				X			X
Overhead duty set by water dew approach		X	X	X		X	X	
Naphtha PA duty set by optimization	X				X			
Naphtha PA to zero duty due to optimization		X	X	X		X	X	X
Naphtha PA duty set by internal liquid wetting								
Kero PA duty set by optimization	X	X	X	X	X	X	X	
Kero PA duty set by internal liquid wetting								X
Diese PA duty set to internal liquid wetting	X	X	X	X	X	X	X	X
Kero stripping steam set by flash spec	X				X			
Kero stripping steam set to minimum rate		X	X	X		X	X	X
Heavy naphtha product as drag stream	X	X	X	X	X	X	X	X
Fired heater before preflash				X				X
Heat integration after preflash		X				X		

Feed	Bakken	Bakken	Bakken	Bakken	Gipps	Gipps	Gipps	Gipps
Configuration	Topping	Stripped	Reboiled	Flashed	Topping	Stripped	Reboiled	Flashed
No Preflash	X				X			
Overhead duty set by overhead temp	X			X	X			X
Overhead duty set by water dew approach		X	X			X	X	
Naphtha PA duty set by optimization		X	X					
Naphtha PA to zero duty due to optimization				X		X	X	X
Naphtha PA duty set by internal liquid wetting	X				X			
Kero PA duty set by optimization	X				X	X	X	
Kero PA duty set by internal liquid wetting		X	X	X				X
Diese PA duty set to internal liquid wetting	X	X	X	X	X	X	X	X
Kero stripping steam set by flash spec	X	X	X	X	X	X	X	X
Kero stripping steam set to minimum rate								
Heavy naphtha product as drag stream	X	X	X	X	X	X	X	X
Fired heater before preflash				X				X
Heat integration after preflash		X				X		

(Shaded) Equipment or process constraints did not limit optimization.

Energy Intensity

- Firing plus steam duty
- What is the value of the energy in steam versus the value of energy in fuel?
- Value differs between plants
- Overall, reboiled configurations favored
- Reboiled configurations limited to naphtha product overhead by bottoms temperature
- Preflash kerosene product operation would require some steam stripping

Energy Intensity

Feed		BCF24	BCF24	BCF24	BCF24	TJL	TJL	TJL	TJL
Config		Topping	Stripped	Reboiled	Flashed	Topping	Stripped	Reboiled	Flashed
Feed API		23.2	23.2	23.2	23.2	31.5	31.5	31.5	31.5
Energy Balance									
Total Firing	Mbtu/hr	194.5	162.4	171.4	211.6	222.6	169.3	183.1	222.4
Total Steam	lb/hr	24522	63688	27018	26869	16655	52369	19661	19522
Total Firing + PF Steam	Mbtu/hr	194.5	199.4	171.4	211.6	222.6	202.2	183.1	222.4
PA Recovery	Mbtu/hr	91.4	43.4	45.8	50.7	152.5	58.6	58.4	72.0
Intensity	btu/bbl distillate	4715	5559	4874	5957	3706	3790	3467	4185
After-PF Integration	Mbtu/hr		70.7				55.6		
Steam/Fired	For equal value		0.195				0.301		
Steam/Fired	No after PF-Integration		-0.915				-0.761		

Feed		Bakken	Bakken	Bakken	Bakken	Gipps	Gipps	Gipps	Gipps
Config		Topping	Stripped	Reboiled	Flashed	Topping	Stripped	Reboiled	Flashed
Feed API		42.0	42.0	42.0	42.0	51.1	51.1	51.1	51.1
Energy Balance									
Total Firing	Mbtu/hr	256.0	168.1	183.2	221.9	217.0	157.5	173.6	219.7
Total Steam	lb/hr	7255	37414	10908	10780	3077	24512	4256	4172
Total Firing + PF Steam	Mbtu/hr	256.0	194.8	183.2	221.9	217.0	178.0	173.6	219.7
PA Recovery	Mbtu/hr	214.9	90.7	89.7	90.0	182.0	85.0	87.9	81.5
Intensity	btu/bbl distillate	3098	2621	2482	2994	2343	1971	1934	2442
After-PF Integration	Mbtu/hr		30.9				17.7		
Steam/Fired	For equal value		0.437				0.697		
Steam/Fired	No after PF-Integration		-0.388				-0.026		

Stripped versus Reboiled Preflash Energy Values

Preheat	BCF24	TJL	Bakken	Gippsland
Additional	0.195	0.301	0.437	0.697
No additional	-0.915	-0.761	-0.388	-0.026

Energy value of steam compared to fuel based on equivalent operating cost for a stripped preflash tower versus a reboiled preflash tower

Conclusions

- Equipment constraints critical to design and operation
  - Equipment performance and limits must be understood
  - Arbitrary equipment limits can have large consequences
  - Effective operation and economics require equipment understanding
- Overall economics favors reboiled units
  - Steam stripping units require low value steam or even negative value steam to be cost effective
- However,
  - Reboiled units limited to heavy naphtha as overhead product; kerosene products require too high a bottoms temperature – oil thermal stability limits operation
- Heat integration can make steam stripping less costly
  - Additional heat integration with preflash tower bottoms
  - More complex integration with vacuum tower
  - Preflash tower stripping-versus reboiling-differs from condensate splitters due to heat integration

Foaming

25

Preflash Stabilizer Foaming

- Foaming a known problem
- Stable foams lead to entrainment if vessel residence time too low
- Configurations here used cyclonic separators to reduce the effect of foam
- More traditional configurations use large residence times
- Foam is one concern that may drive selection to a preflash configuration – large holdup volumes are easier to add
- Some alternatives can provide foam holdup volume in a steam stripped or reboiled condensate splitter
- Preflash with steam injection – boosts preflash yields
- Split towers – adds foam holdup volume
- Foam drum – adds foam holdup volume

CH2MHILL. 26

Recommended Selection Strategy

27

Selection Strategy

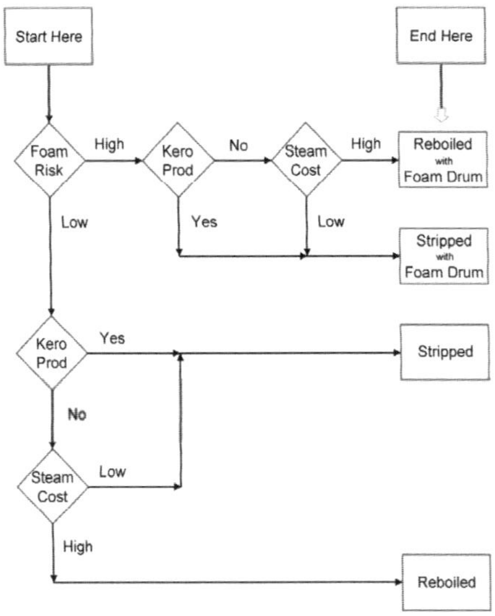

CH2MHILL.

28

Questions ?

29

Revisions

- 2015-03-08: Rev 0

CH2MHILL.

30

402154 Coking Mechanism in Refinery Fractionator Wash Beds

Abstract:

This paper will discuss a novel theory of the coking mechanism in a refinery fractionator wash bed. The wash bed of a refinery fractionator is prone to coking/fouling due to the low liquid rates along with high vapor rates and high temperatures. Generally, the wetting rate at the top of the wash bed is minimized to prevent high-value product loss, but lower wetting rates lead directly to coke formation in the bed. So the refinery works to balance the economics of product recovery versus unscheduled downtime to replace a coked bed. Marathon has used TRACERCO Diagnostics™ ThruVision technology to help manage the wash bed life while maximizing product recovery during the normal turnaround cycle. This technology provides a detailed density map at a specific vertical elevation that can pin point specific areas of liquid maldistribution or solids build-up. During a turnaround cycle, several towers have experienced heater trips that have led to step changes in the wash bed density seen by the ThruVision technology and ΔP. A review of this data from several refinery fractionators has guided the creation of the theory on how the coke is formed in a refinery fractionator wash bed.

402162 How Does a Collector Tray Leak Impact Column Operation?

Abstract:

Collector trays are commonly used in distillation columns to provide positive material balance control. They are designed with utmost care so that liquid can collect and flow without hydraulic restriction and without leaking. Controlling liquid leakage from a collector tray is paramount to ensuring adequate column heat and material balance. Many heavy oil fractionators, including FCC Main Fractionators, use collectors for material balance.

Most operating and engineering companies require tight leakage criteria for collector trays during installation. Engineers often envision that a leak would result in a discreet, predictable stream of liquid from the collector tray, falling neatly onto the fractionation device below. The lost product would instead contribute additional reflux to the section below – a neat and convenient story – while slightly un-optimized, would result in limited operational difficulty. Along with many in industry, the authors held this belief before troubleshooting the two case studies discussed in this paper.

The authors will present the results from two FCC Main Column troubleshooting cases. The first case is from a 100 MBD FCC Main Column that was having trouble yielding LCO at design rates. The second case is from a 30 MBD FCC Main Column that had severe instability above a certain collector tray liquid level where leakage commenced. Both towers exhibited leaking LCO collector trays, but drastically different resultant unit performances due to the type of internals below the collector tray (trays vs packing).

402155 Prevent Damage in FCC Main Column Slurry Pumparound Zones with Proper Design and Operation

Abstract:

Severe service grid packing has become widely used in FCC Main Fractionator slurry pumparound zones to improve capacity and heat removal efficiency. Although most units operate without any issues or mechanical damage, there have been reports of isolated, but repeated, incidents involving severe damage to the internals in some Main Fractionators. The analysis of one case where significant damage was observed led to a determination of the most probable cause of the damage and resulted in a solution to prevent future damage. Proper unit operation in conjunction with a robust mechanical design of the packed bed internals is recommended as the best method to ensure reliable operation in this very severe application.

397295 Maximize Performance & Run Lenght of a CDU Main Fractionator with Advanced Mass Transfer Components

Abstract:

Maximizing capacity and flexibility, improving efficiency, and increasing run length of the Main Fractionator is of the esence for a Crude Distillation Unit, particularly when processing different types of Crude, including heavy and sour feedstock. Among other steps, these goals can be achieved only with a proper combination of differen types of mass transfer components, suitable not only for high performance but also for great fouling and corrosion resistance, particularly at the top section of the column.

Folowing the great success of the revamp implemented 5 yesars ago, a similar retrofitting has been implemented in another twin Unit of the same Refinery. This paper will describe in details the revamp of the Unit, including the modifications of the Main Fractionator, and the upgrading pf the process control.

CH2MHILL.

Condensate Splitting – What are the Options?

Andrew W. Sloley – CH2M HILL, Bellingham, WA

Prepared for Presentation at the AIChE 2015 Spring Meeting, Austin, TX, 26-30 April, 2015

Condensates Examined

2

Condensate

- Light oil with an API gravity of more than 45°
- Most streams named as a condensate have an API gravity of 49°+
- Source
  - Well production
  - Retrograde condensate from pipeline systems
- Production increasing due to production from tight oils
- Produces a large amount of light-ends and light naphtha product
- Loads up refinery atmospheric tower
- Possibility of asphaltene instability when mixing with heavier oils
- Refinery light-ends containment
  - Too much material with a high RVP (propane and n-butane)
  - Limited sales and ability to blend into gasoline

CH2MHILL. 3

Condensate Splitting

- Possibility to split light naphtha through heavy naphtha from condensate
- Generates very high RVP overhead product
- Overhead exportable for gasoline
  - Special shipping and handling
  - Or stabilize before shipping
- Bottoms product contains middle distillate and heavier for domestic use
- Export gasoline components
- Use middle distillate internally

CH2MHILL. 4

Condensates

- Cover range of commercial interest
- Cover Reid Vapor Pressure (RVP) range
  - Stabilized condensates have low RVP
  - Unstabilized or partially stabilized condensates have high RVP
- Cover API° gravity range being traded
  - High API° gravity = low specific gravity
  - High API° gravity = high light naphtha yields – common perception
- Assay quality must include sufficient information on light-end to match RVP

CH2MHILL. 5

Condensates Examined

Condensate	Density		RVP		Origin
	API°	SG	psia	kPa	
Åsgard	50.6	0.7770	9.32	64.2	Norway
Griffin	55.0	0.7587	5.40	37.2	Australia
Sleipner Kårstø	57.5	0.7487	10.3	70.7	Norway
Northwest Shelf	59.5	0.7408	8.8	60.7	Australia
F3FB	63.6	0.7253	14.1	97.2	Netherlands

CH2MHILL. 6

Selection Diagram

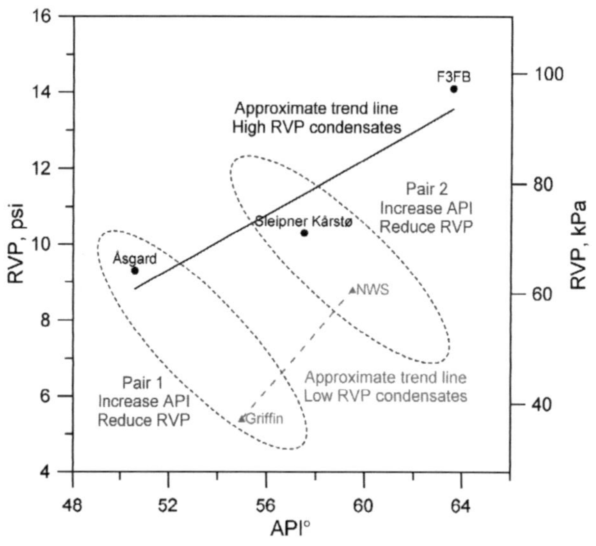

CH2MHILL.

7

Configuration Options

8

Configuration Options and Possible Operations

- Steam Stripping
  - Preheat feed to flash
  - Recover light material from bottoms by vaporization with reduced partial pressure
- Reboiled
  - Preheat feed to flash
  - Recover light material from bottoms by fractionation
- Preflash
  - Preheat feed to flash
- Operations
  - Light naphtha overhead
  - Heavy naphtha overhead
  - Others possible

CH2MHILL. 9

Steam Stripping

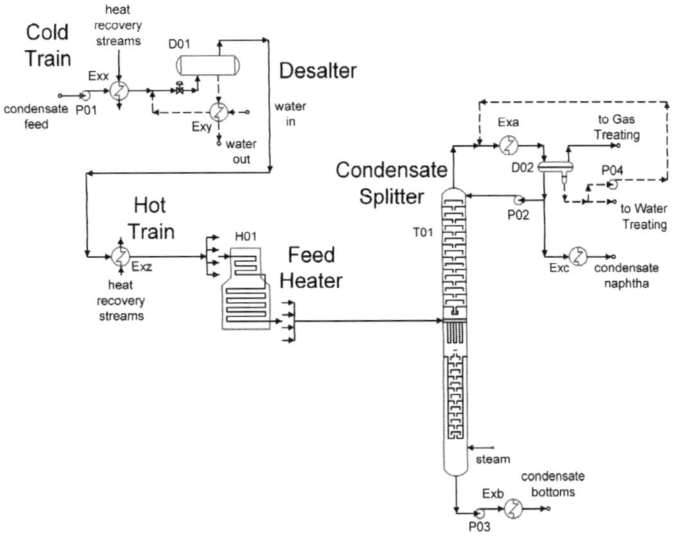

CH2MHILL. 10

Reboiled

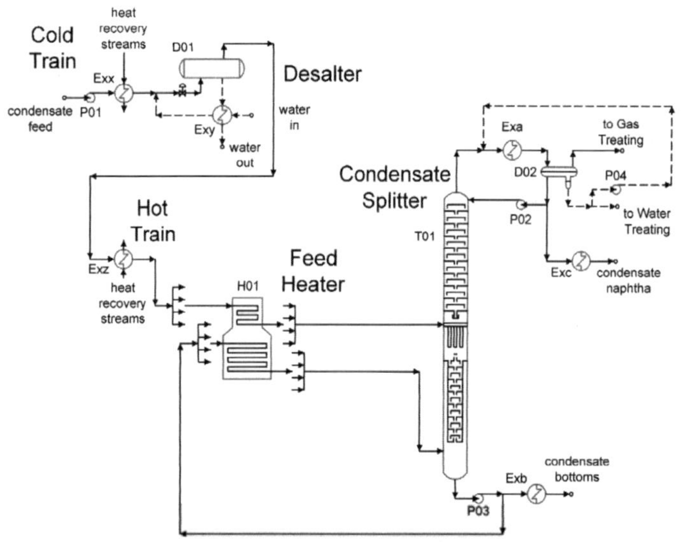

CH2MHILL.

11

Preflash

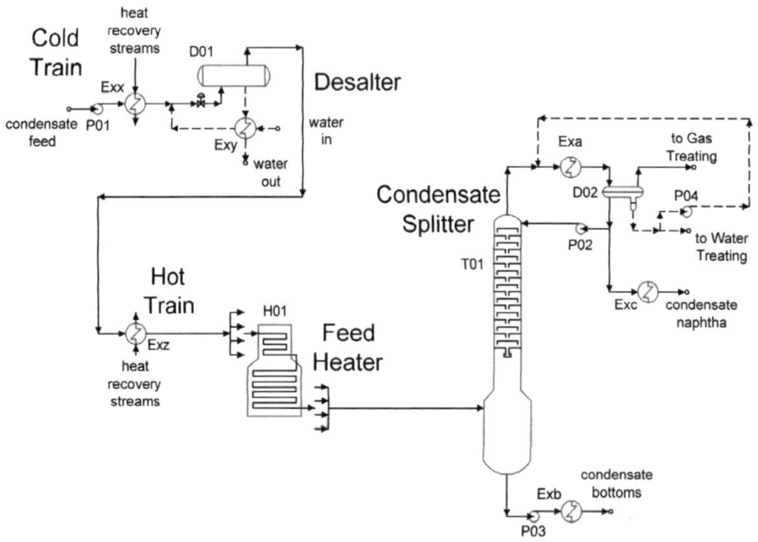

CH2MHILL.

12

Comparisons

13

Comparisons Made

- Five condensates
- Three configurations
- Two operations
- Low-cutpoint operation – 230°F (110°C) overhead D86 95 vol %
  - Deals with RVP containment at the refinery
- High-cutpoint operation – 350°F (177°C) overhead D86 95 vol %
  - Conserves maximum middle distillate (kerosene+diesel) in the bottoms
- 30 total units evaluated

CH2MHILL.

14

Evaluation Basis – Key Points

- Independent unit
- Maximum heat integration to 40°F (22.2°C) approach temperature
- Optimized heat input in bottoms versus feed preheat for fired heaters
- Stripped unit as base configuration
- Yield volumes matched to base configuration when possible
- Equipment constraints for good practice imposed on
  - Minimum exchanger duty
  - Fractionation liquid and vapor rates
  - Fractionation temperature profiles to avoid corrosion
  - Fired heat outlet temperatures and feed vaporization

CH2MHILL. 15

Fractionation Stages versus Reflux

- Preliminary work showed very strong driver towards maximum trays
  - Maximum trays = minimum reflux
  - Minimum reflux = minimum energy downgrade
  - True even with low energy price level of $4 per Mbtu gas price (3.79 $ GJ)
  - More expensive energy makes this driver even more powerful
- 15 theoretical stages used in all rectification sections
- Sensitivity analysis showed 15 stages best overall choice
- Same number of stages used to generate TBP data via D2892 method

CH2MHILL. 16

Stripping Trays versus Reboiled Trays

- Same physical number of trays used in the steam stripping configuration versus reboiled configuration
- This approach gives an approximate equal investment analysis
- Steam stripping trays have much lower efficiency
- Result is
  - Steam stripping sections – 3 theoretical stages
  - Reboiled stripping sections – 9 theoretical stages

CH2MHILL.

17

Results and Conclusions

18

Constraints

- Most cases constrained by equipment limitations rather than optimum thermodynamics or economics of recovery versus energy
- 30 cases studies
  - 18 with one or more equipment related constraints
  - 5 with one or more equipment related constraints PLUS process constraints
  - 7 reached an optimum operation

Constraint Summary

Feed	Asgard	Asgard	Asgard	Griffin	Griffin	Griffin	SleipK	SleipK	SleipK	NWS	NWS	NWS	F3FB	F3FB	F3FB
Operation	Low Cut	Low Cut	Low Cut	Low Cut	Low Cut	Low Cut	Low Cut	Low Cut	Low Cut	Low Cut	Low Cut	Low Cut	Low Cut	Low Cut	Low Cut
Configuration	Strip	Reboil	Preflash	Strip	Reboil	Preflash	Strip	Reboil	Preflash	Strip	Reboil	Preflash	Strip	Reboil	Preflash
Naph Yield	Base	X	X	Base		X	Base	X	X	Base	X	X	Base	X	X
Top Liquid - Feed Q	X			X			X	X		X	X		X	X	
Top Liquid - Rebl Q															
Bottom Vapor					X										
Water Condensation				X											
Preheat Config		X	X	X	X					X					
Feed Vaporization															

Feed	Asgard	Asgard	Asgard	Griffin	Griffin	Griffin	SleipK	SleipK	SleipK	NWS	NWS	NWS	F3FB	F3FB	F3FB
Operation	High Cut	High Cut	High Cut	High Cut	High Cut	High Cut	High Cut	High Cut	High Cut	High Cut	High Cut	High Cut	High Cut	High Cut	High Cut
Configuration	Strip	Reboil	Preflash	Strip	Reboil	Preflash	Strip	Reboil	Preflash	Strip	Reboil	Preflash	Strip	Reboil	Preflash
Naph Yield	Base	X	X	Base		X	Base	X		Base	X		Base	X	
Top Liquid - Feed Q	X			X			X			X	X		X	X	
Top Liquid - Rebl Q					X										
Bottom Vapor		X													
Water Condensation															
Preheat Config				X						X					
Feed Vaporization									X			X			X

Unconstrained or meets process requirements
Constrained
Not applicable
Fails to meet process requirements

Energy Consumption

- All low-cut operations required low steam prices for steam stripping to be attractive
- The lighter the condensate, the more attractive the reboiled operation
- All high-cut operations, except one, required negative steam prices for stripping steam to be attractive
- The heaviest condensate showed stripping steam attractive
  - High bottoms rate
  - Duty required to meet minimum vapor loads in bottom trays
  - This limit could be worked around with more complex tray designs
- Overall, reboiled configurations favored
- Reboiled configurations limited to naphtha product overhead by bottoms temperature
- Kerosene cutpoint operation would require some steam stripping

Energy Summary

Configuration	Operation		Condensate				
Fired duty in feed heater plus reboiler plus steam duty							
	Low-cutpoint		Asgard	Griffin	SleipK	NWS	F3FB
Stripping		btu/bbl	4300	6231	2703	2824	2846
Reboiled		btu/bbl	3497	7101	2431	2421	2390
Preflash		btu/bbl	5623	6576	2940	3018	2916
	High-cutpoint						
Stripping		btu/bbl	2512	2419	2248	2133	2120
Reboiled		btu/bbl	2991	1914	2070	1971	1961
Preflash		btu/bbl	3730	3029	2464	2249	2391
Energy comparison - Shifting from stripping to reboiling							
	Low-cutpoint						
Firing change		btu/bbl	1,081	2,210	279	277	219
Steam change		btu/bbl	-1,884	-1,340	-551	-680	-675
Breakeven steam:firing value		fraction	0.57	1.65	0.51	0.41	0.32
	High-cutpoint						
Firing change		btu/bbl	664	-217	-89	-70	-52
Steam change		btu/bbl	-185	-288	-89	-93	-107
Breakeven steam:firing value		fraction	3.60	-0.75	-1.00	-0.75	-0.48

RVP versus Yield

- Small amounts of light-end material create significant vaporization changes in tower feed
- This can dramatically shift yields
- Swings in light-end content will require control and equipment flexibility

CH2MHILL. 23

RVP and Yield Summary (Reboiled Splitter)

Operation		Condensate				
		Åsgard	Griffin	SleipnerK	NWS	F3FB
Feed						
API°		50.6	55.0	57.5	59.5	63.6
RVP, psia		9.32	5.40	10.3	8.8	14.1
SG		0.7770	0.7587	0.7487	0.7408	0.7253
RVP, kPa		62.4	37.2	70.7	60.7	97.2
Naphtha						
Low-cut	bpd	20,706	9,238	44,803	41,223	43,759
Low-cut	cmpd	3,292	1,469	7,124	6,554	6,958

CH2MHILL. 24

Conclusions

- Equipment constraints critical to design and operation
  - Equipment performance and limits must be understood
  - Arbitrary equipment limits can have large consequences
  - Effective operation and economics require equipment understanding
- Overall economics favors reboiled units
  - Steam stripping units require low value steam or even negative value steam to be cost effective
  - Reboiled units limited to heavy naphtha as overhead product; kerosene products require too high a bottoms temperature – oil thermal stability limits operation
- Light-ends content significantly effects yields
  - Small amounts of light-ends significantly change feed vaporization

CH2MHILL. 25

Foaming

26

Condensate Stabilizer Foaming

- Foaming a known problem with condensate and light crude oil
- Stable foams lead to entrainment if vessel residence time too low
- Base configurations used cyclonic separators to reduce the effect of foam
- More traditional configurations use large residence times
- Foam is one concern that may drive selection to a preflash configuration – large holdup volumes are easier to add
- Some alternatives can provide foam holdup volume in a steam stripped or reboiled condensate splitter
- Preflash with steam injection – boosts preflash yields
- Split towers – adds foam holdup volume
- Foam drum – adds foam holdup volume

CH2MHILL. 27

Preflash with Steam Injection

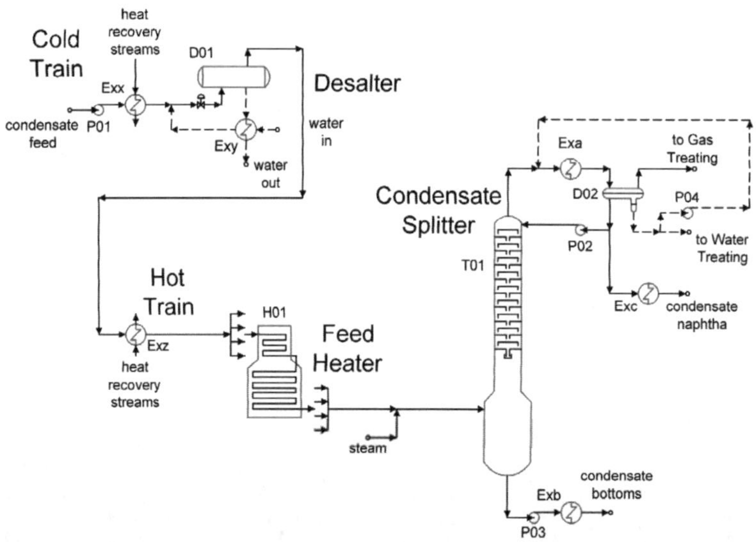

CH2MHILL. 28

Reboiled with Split Tower

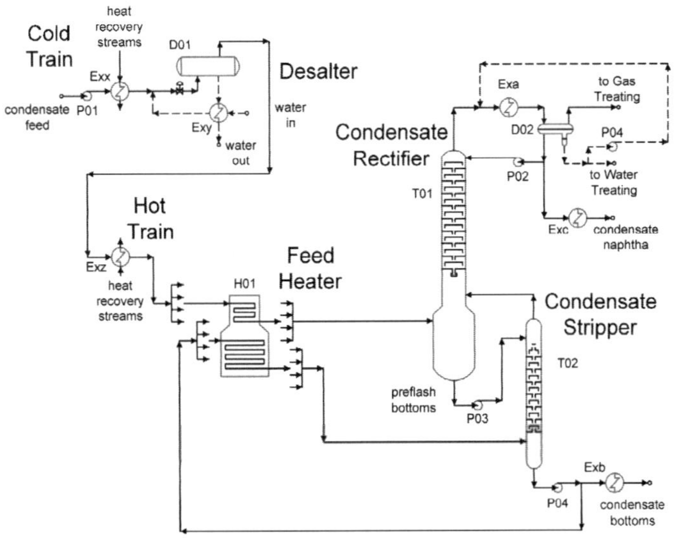

CH2MHILL.

29

Reboiled with Foam Drum

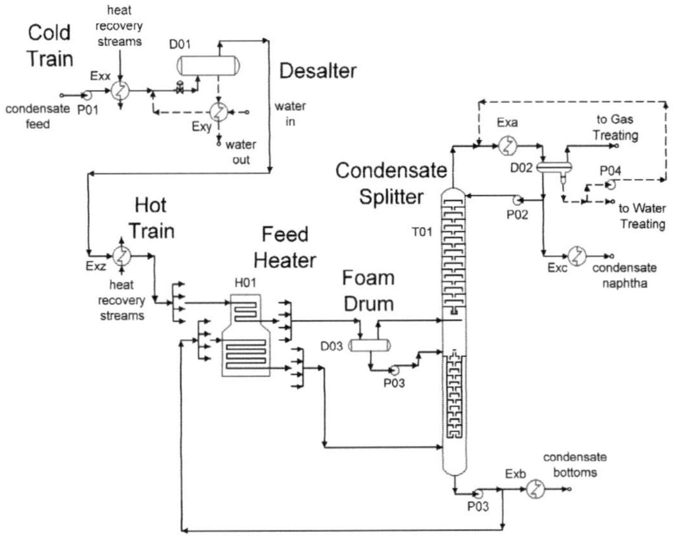

CH2MHILL.

30

Recommended Selection Strategy

31

Selection Strategy

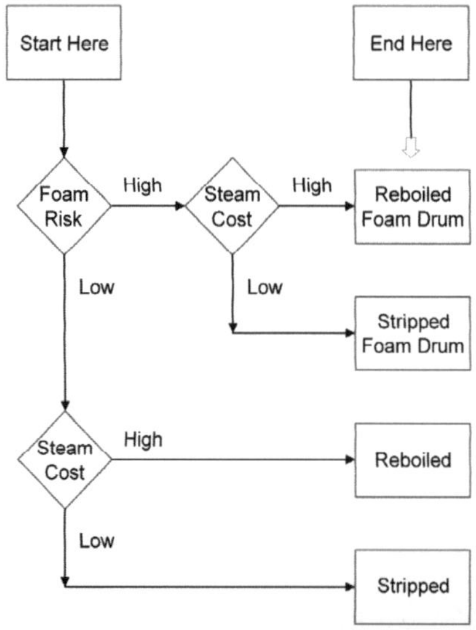

CH2MHILL.

32

Questions ?

33

Revisions

- 2015-02-10: Rev 0
- 2015-03-08: Rev 1 – typographic corrections

CH2MHILL.

34

400211 Slippery Slopes: How Should We Calculate the Equilibrium Slope in Multicomponent Distillation?

Abstract:

The calculation of tray efficiencies the HETP of packing requires, among other things, the slope of the equilibrium line. For binary systems the slope is unambiguously defined (although not always rigorously calculated). However, there is no unique definition of the slope for multicomponent systems because the vapor liquid equilibria is properly represented by a surface (in hyperspace if there are more than 3 compounds).

Thus, for multicomponent systems any pseudo-binary method for estimating tray efficiencies and HETP must rely, in part, on an approximate method of estimating the slope. But which approximation is best? And what does "best" mean when no rigorous definition is even possible?

Thus, our first purpose here is to demonstrate that different approximations can lead to dramatically different estimates of the equilibrium slope.

The second goal of this presentation is to investigate the extent to which the choice of slope estimation method matters in the design of distillation processes.

Simulation of Batch Distillation with a Varying Alpha (α): The Ethanol-Water and Methanol-Water Systems at 1 Atm

Robert G. Kunz

RGK Environmental Consulting, L.L.C.
Hillsborough, North Carolina

Paper 99b for Presentation at

2015 AIChE Spring Meeting

April 26-30, 2015

Austin, Texas

Abstract

In a previous presentation **[1]**, a newly developed technique was described that enables analytical integration of the Rayleigh Equation for batch distillation with a varying relative volatility (α). Heretofore, integration in closed analytical form has required that α be assumed constant. This new technique produces a continuous analytical function for the Rayleigh Equation integral for a varying α and reduces algebraically to the traditional function when α is constant. Detailed derivations are contained in the cited presentation.

Both the traditional expression at constant α and the new equation were evaluated against numerical integration via Simpson's Rule for a range of typical values of constant α, for a number of ideal and non-ideal binary vapor-liquid equilibrium (VLE) systems with varying α, and for an illustrative example contained in a chemical engineering textbook. The method was then utilized successfully in a computer-simulated batch distillation of an ethanol-water solution at 1 atm pressure. Results compared favorably with expected values.

The present paper reviews the ethanol-water case and extends the simulation procedure to methanol-water. Batch distillation of these systems at atmospheric pressure is a popular student laboratory exercise in colleges and universities throughout the world.

A summary of VLE relationships for ideal and non-ideal systems plus temperature-dependent methods for analysis of constituent composition specific to both the ethanol-water and methanol-water systems are discussed in separate appendices. These latter include, among others, estimates of liquid density for alcohol-water mixtures extended beyond the range of published data and refractive index measurements assembled from various sources.

The paper consists of new material supported by background information tutorial in nature.

Keywords. Alcohol "Proof"; Boiling Point Curves; Computer Simulation, Density, Refractive Indices, Gas Chromatography, Flash Points, and Auto-ignition Temperatures for Ethanol-Water and Methanol-Water Solutions; Numerical Integration; Rayleigh Equation; Simpson's Rule; Systems: Ethanol-Water, Methanol-Water, and Methanol-Ethanol; Vapor-Liquid Equilibrium.

Disclaimer
("Some restrictions apply; batteries not included; your mileage may vary")

The following caveat applies: The information contained herein is offered in good faith but without guarantee, warranty, or representation of any kind (expressed or implied) as to its usefulness, correctness, completeness, or fitness for any particular purpose. The user assumes all risk for its implementation and should seek independent professional verification of its accuracy. The author assumes no responsibility and shall not be liable for any loss of profit nor any special, incidental, consequential, or other damages which may result from the use of any of the information contained in this presentation, be it oral or written. Any statements concerning design, construction, operation, what constitutes regulatory compliance, and/or how to achieve such compliance should not be construed as recommendations on the part of the author and/or his organization.

Introduction

Distillation is a separation technique utilizing the difference in composition of a liquid and the vapor in equilibrium with that liquid **[2,3]**. Simple, or differential, distillation employs a single equilibrium stage to effect that separation. A batch distillation operates on an unreplenished charge of material being boiled off from a still pot and then condensed as product in another vessel. It is an unsteady-state process, where the amount and composition of feed and product vary with time. In a two-component (binary) system, the charge of feed material boils away and gradually becomes richer in the less volatile/higher-boiling/ heavy component. The condensed vapor product (distillate) increases in volume and is enhanced in the more volatile/lower-boiling/light component. This paper discusses the simple, differential batch distillation of a two-component (binary) system, in which none of the condensed vapor is returned, or refluxed, to the still pot. A typical laboratory setup is shown in Figure 1.

Simple batch distillation is not practiced extensively in industry. It is used commercially for separating small quantities of high-value chemicals, when high purities are not required, as a preliminary step to be followed by further processing, for separations where the equipment requires frequent cleanout, or where the feed mixture is very easy to separate **[2,3]**. A rule of thumb for ease of separation is that the feed components differ in atmospheric boiling point by at least 25°C [**4**, pp.167-168], 30-40°C [**5**, p.61], or 75°C [**6**, p.89; **7,** p.89], depending on the source. Whatever the precise numerical value, the intent is that the lighter component be much more easily vaporized in preference to the heavier component in order to distill a decent amount of high-purity product before too much of the feed material boils away.

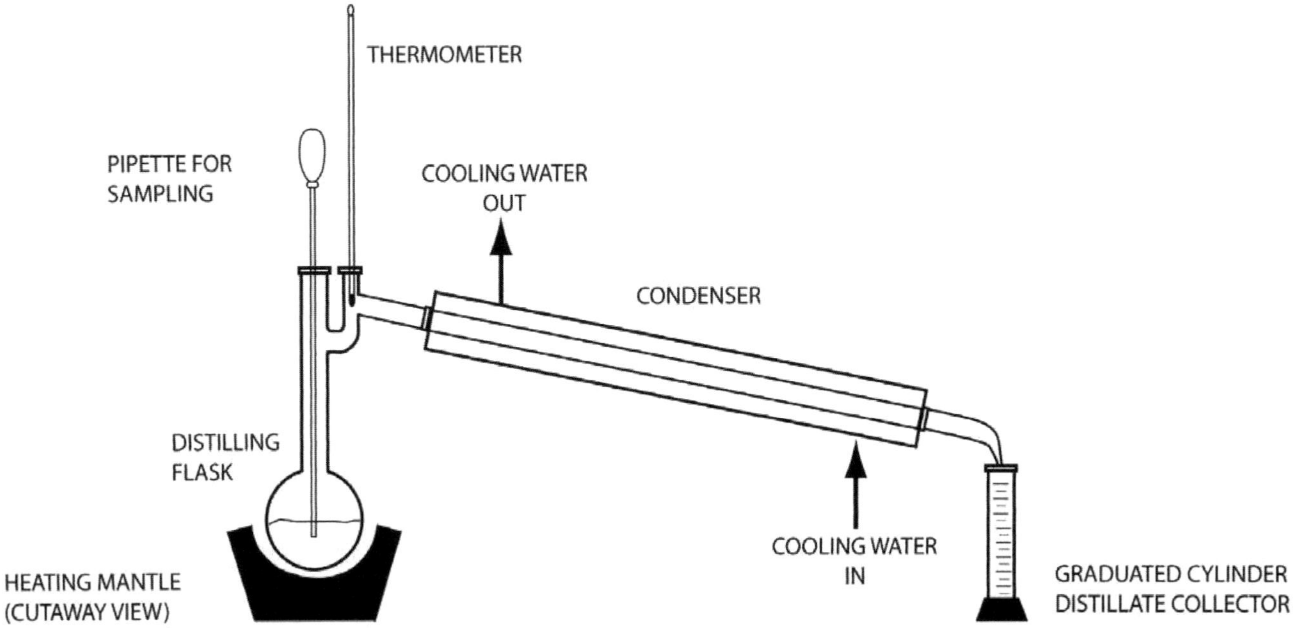

Figure 1. Typical Laboratory Setup for Simple, Differential Batch Distillation

The Rayleigh Equation

Simple batch distillation without reflux is described by the Rayleigh Equation, first advanced by Lord Rayleigh in 1902 **[8]**. A step-by-step derivation [**2**; **9**, p.580-581; **10**; **11**] is contained in a previous presentation.

$$\ln (W/F) = \int_{x_F}^{x_W} dx/(y - x) \tag{1}$$

where

F	=	total moles of feed in the still pot/distilling flask at starting time (t) = 0
W	=	total moles left in still pot (bottoms) at any time
D	=	total moles boiled off and condensed into the distillate at the same point in time
x	=	mol fraction of the more volatile component (MVC) in the liquid phase
y	=	mol fraction of the MVC in the vapor phase in equilibrium with x

Other quantities are determined by material balance or a component mass balance:

$$F = W + D \tag{2}$$

$$Fx_F = Wx_W + Dx_D \tag{3}$$

$$x_D = (Fx_F - Wx_W) / (F - W) \tag{4}$$

For the first drop of distillate, when $W = F$ and $x_W = x_F$, the expression in Equation (4) becomes mathematically indeterminate. and x_D is provided by the x-y equilibrium curve, as is the instantaneous composition of the distillate at every moment during the distillation.

The Rayleigh Equation appears simple; yet its evaluation is somewhat involved. The integrand is not a simple function of x: at constant pressure, every value of x corresponds to a different temperature; except for the most ideal systems, y is not related to x in a simple manner. Important variables are temperature (t), total pressure (P_T), partial pressure (p), vapor pressure (p^{vap}), liquid-phase (x) and vapor-phase (y) mol fractions, plus so-called activity coefficients (γ's) when a given system cannot be considered to be ideal. Equations for low-pressure binary vapor-liquid equilibrium (VLE) are summarized in Appendix A.

The standard procedure is to evaluate the integral numerically or graphically,* or to come up with a suitable substitution in the hopes of producing an analytical solution in closed mathematical form. Numerical integration involves repeatedly fitting the numerical values of the integrand with a mathematical function over short ranges, integrating those functions mathematically, and then summing. This numerical fitting and summing process results in a series of discrete values, rather than a continuous function.

One analytical approach makes use of a parameter known as the *relative volatility*, denoted by the symbol *alpha* (α). It is defined as the ratio of the vapor-phase mol fraction of the more volatile component, denoted as *component 1*, to its liquid-phase mol fraction, this ratio being divided by a similar ratio for the less volatile *component 2*. For a successful separation by distillation, α must be greater than 1.0. In the case of a two component system

$$\alpha = (y_1/x_1) \,/\, (y_2/x_2) = (y_1/x_1) \,/\, [(1 - y_1) \,/\, (1 - x_1)] \qquad (5)$$

since $y_2 = (1 - y_1)$ and $x_2 = (1 - x_1)$.

Rearranging Equation (8), one obtains for $1 / (y_1 - x_1)$ to be substituted in Equation (1).

$$\frac{1}{(y_1 - x_1)} = \frac{1 + (\alpha - 1)\, x_1}{(\alpha - 1)\, x_1\, (1 - x_1)} = \frac{1}{(\alpha - 1)\, x_1\, (1 - x_1)} + \frac{1}{(1 - x_1)} \qquad (6)$$

The relative volatility (α) is evaluated at constant total pressure for any temperature or the corresponding value of composition at every point along the vapor-liquid equilibrium (VLE) x-y curve from a ratio of vapor pressures and, if necessary, activity coefficients:

$$\alpha = (y_1/x_1) \,/\, (y_2/x_2) = (p_1^{vap}\, \gamma_1) \,/\, (p_2^{vap}\, \gamma_2) \qquad (7)$$

The traditional approach has been to assume α to be constant and integrate Equation (6) to obtain Equation (8) below [**14**, #40, #29]:

---

* In previous times, in addition to counting boxes under the curve of the integrand or using a planimeter device [**12**], the curve was drawn carefully on high quality graph paper of uniform density and then was weighed and calibrated against a rectangular section of that same graph paper. This method is noted in Ref. [**13**, p.75].

$$\ln (W/F) = \frac{1}{(\alpha - 1)} \left\{ \alpha \ln [(1 - x_F)/(1 - x_W)] + \ln (x_W/x_F) \right\} \tag{8}$$

The percent of original charge remaining in the still pot is then given by:

$$100\,(W/F) = 100 \exp [\ln (W/F)] \tag{9}$$

However, even for the most ideal of systems, relative volatility is not exactly constant over the entire range of x.

A New Twist

Nevertheless, alpha need not necessarily be constant to obtain a closed-form analytical solution to Equation (1) as long as the factor $1/(\alpha - 1)$ can be fitted as a function of x over the range of interest, for example:

$$1 / (\alpha - 1) = a\,x^2 + b\,x + c \tag{10}$$

This is much the same technique as the expression of gaseous heat capacities by an empirical power series in temperature.

With $1/(\alpha - 1)$ so represented, substitution in Equation (6) produces Equations (11) and (12), which can then be integrated analytically term by term [**14**, #32, #29, #40]:

$$\ln (W/F) = \int_{x_F}^{x_W} \frac{(a\,x_1^2 + b\,x_1 + c)\,dx_1}{x_1\,(1 - x_1)} + \int_{x_F}^{x_W} \frac{dx_1}{(1 - x_1)} \tag{11}$$

$$\ln (W/F) = \int_{x_F}^{x_W} \frac{a\,x_1\,dx_1}{(1 - x_1)} + \int_{x_F}^{x_W} \frac{(1 + b)\,dx_1}{(1 - x_1)} + \int_{x_F}^{x_W} \frac{c\,dx_1}{x_1(1 - x_1)} \tag{12}$$

When the smoke clears and the dust settles, one is left with the following:

$$\ln (W/F) = a\,(x_F - x_W) + (1 + a + b + c) \ln [(1 - x_F) / (1 - x_W)] + c \ln [\,x_W / x_F\,] \tag{13}$$

The only source of error is the goodness of fit of the quadratic function in Equation (10). When α is constant for all values of composition, a = b = 0, c = 1/ (α – 1) identically, and Equation (23) reduces to Equation (13), the solution of the Rayleigh Equation with α constant. If necessary, an x^3 term can be added to Equation (10) but adds more terms to Equation (13).

Testing of the New Function against Numerical Integration

Simpson's Rule. Both the traditional expression at constant α and the new equation were evaluated against numerical integration via Simpson's Rule for a broad range of typical values of constant α, for a number of ideal and non-ideal binary vapor-liquid equilibrium (VLE) systems with varying α, and for a illustrative example contained in a chemical engineering textbook. A summary is given below; details have been presented previously **[1]**.

It has been found that repeated application of Simpson's Rule in steps of 0.02 and an interval size of 0.01 mol fraction reproduces the exact integral at constant α for ln (W/F), even when exponentiated, to a degree better than can be measured experimentally. This situation applies down to a mol fraction bordering on 0.02. Narrowing the interval size at the lower end of the mol fraction range, where the integrand begins to increase rapidly, maintains the accuracy of the numerical integration even further. Eventually, however, the integrand grows without bounds as mol fraction remaining in the still pot approaches zero.

Numerous real vapor-liquid equilibrium systems were then chosen to test how well Equation (23) works in practice over the range of α from $x_1 = 0.02$ to $x_1 = 0.5$ when alpha is not exactly constant. Here, the previously validated Simpson's Rule integration is the standard, using the same interval size found suitable for the numerical integration with constant α. The overall conclusion is that agreement of the newly derived Equation (23) with numerical integration is as good as when α is constant. In addition, one has a continuous function in Equation (23) as opposed to numerical integration, which is valid only at discrete points.

Illustrative Example. The technique produced virtually identical results when tested against an illustrative example presented in McCabe and Smith **[11]** for batch/differential distillation of a 50/50 mol percent benzene-toluene solution at 1 atm pressure with 1/α taken as constant at 0.41. Calculated values for temperature and benzene compositions at various locations were plotted in the illustrative example against the mol fraction of distillate / charge distilled (D/F).

Testing of the New Function in Simulated Laboratory Experiment. Finally, to demonstrate the ease of use of the newly developed technique, batch distillation of a pure ethanol-water solution was simulated at a standard atmospheric pressure of 760 mmHg (1.013 bar). The methanol-water system has been added for discussion here. Results compare favorably with expected values. Both are favorite student laboratory exercises and constitute the main subject for the present paper.

Simulation of Laboratory Experiments

Experimental. An initial 200-mL charge of a pure ethanol-water or methanol-water solution, 20% alcohol by volume at 60 °F (15.56 °C) in an apparatus similar to Figure 1 at standard atmospheric pressure (760 mmHg, 1.013 bar) is typical **[6,7,15]**. This corresponds to 16 wt % approximately, and ~ 0.07-0.1 mol fraction alcohol (Tables 1 and 2). Volume of the distilling flask is 500 millimeters (mL). Boiling stones are added to prevent bumping **[4**, pp.146,170**]**.

Samples of distillate and perhaps still bottoms are taken as every 5 mL of distillate is collected, and temperature is read at such times. The experiment is over when 50 mL of distillate has been collected. Sometimes the total collected distillate is subjected to further purification in a second step using a separate refluxed fractionating column. This experiment has also been conducted using denatured alcohol and water, fermented sugar solutions, and wine being distilled into brandy.

The student is cautioned to wait to read the temperature until droplets condense on the thermometer bulb **[4**, p.172]. The experiment then proceeds drop by drop. The recommended distilling rate varies, but numbers in the range of 2 **[6,7]** to 10 **[4**, p.171**]** drops per minute are stated. At an average 6 drops per minute (1 drop every 10 seconds), and 20 drops to the millimeter, collection of 50 mL would take just under 3 hours once equilibrium conditions are achieved, exclusive of analytical time. This would allow completion of the first step of the experiment within a typical standard laboratory period.

Quantitative analysis is done by refractive index, density, or possibly gas chromatography (Appendix B). Whatever the method of analysis, it is good practice to prepare one's own calibration curve(s) using the materials and analytical equipment employed rather than relying exclusively on literature data. A qualitative analysis for ethanol in the distillate is often conducted by attempting to ignite several drops of solution on a watch glass, since ethanol and its aqueous solutions greater than about 50 % by volume are combustible at room temperature **[16]**. This technique would also work for methanol, whose aqueous solutions containing ~ 40 % or more by volume also ignite **[17]**. (Refer again to Appendix B.)

Simulation. A computer simulation of the laboratory experiments described above was conducted using a personal computer (PC) spreadsheet. The simulated experiments are valid only for a binary mixture of pure alcohol and water, with an x-y VLE curve as in Figure A2 or Figure A3. It does not apply quantitatively to denatured ethanol containing methanol and other ingredients, fermented sugar solutions, or wine being distilled into brandy.

The simulation begins when equilibrium becomes established as the temperature in the laboratory experiment stabilizes and the first drop of liquid falls into the receiving vessel. The simulation also assumes that the distillation takes place slowly enough so as to maintain true vapor-liquid equilibrium with a drop of liquid condensed on the thermometer at all times.

Results from the simulations are summarized in Tables 1 and 2 for the first 50 mL of distillate collected at 20 °C. Temperature, amount of bottoms remaining and distillate formed, plus composition of bottoms and distillate are recorded at every 5 mL. There are two ways to conduct such an experiment – save each 5-mL increment separately for individual analysis, or let the distillate accumulate as a composite sample. Values are listed both ways in the tables.

Gram moles of bottoms (W) and mol fraction of alcohol in the bottoms (x_W) are related by Equation (13), using constants for $1/(\alpha - 1)$ from Equation (10) fitted in the narrow ranges of interest for x_W (a = 1.73317, b = 0.511067, c = 0.103302 for ethanol and a = 0.449813, b = 0.315463, c = 0.146141 for methanol). These constants provide a better fit for $1/(\alpha - 1)$ over the limited range of x_W in this "experiment" rather than the coefficients listed in Ref. **[1]** covering a larger span of x_W.

Moles of distillate (D) and mol fraction in the distillate (x_D) are computed from material balances [Equations (2) and (3), rearranged]. Moles are converted to grams (gm) using a molecular weight of 46.07 for ethanol, 32.04 for methanol, and 18.02 for water from Table A1:

$gm\ C_2H_5OH = 46.07 \times mol\ C_2H_5OH,\quad gm\ CH_3OH = 32.04 \times mol\ CH_3OH,\quad gm\ H_2O = 18.02 \times H_2O$ (14)

Mol fraction (m.f.) of ethanol or methanol is converted to wt % by:

Ethanol: $wt\ \% = 100 \times (m.f.)\ (46.07)\ /\ [(m.f.)\ (46.07) + (1 - m.f.)\ (18.02)]$ (15)

Methanol: $wt\ \% = 100 \times (m.f.)\ (32.04)\ /\ [(m.f.)\ (32.04) + (1 - m.f.)\ (18.02)]$ (16)

Mass is converted to volume by means of density at known temperature and composition, either at an assumed laboratory temperature of 20 °C or as estimated at the temperature of a distillate sample or that of the solution remaining in the distilling flask (Appendix B). Tables in Ref. [**18**, Tables 6.22 and 6.36], interpolated if necessary, relate wt % and % by volume for ethanol at the standard temperature of 60 °F (15.56 °C). For methanol, vol % at 60 °F was similarly determined from the density table in Perry's Handbook [**19**] and Equation (B2).

Graphical Results

Results of the simulation, extended past the entries in Tables 1 and 2, are plotted in Figures 2 and 3 following the same functional form suggested in the illustrative example for benzene-toluene in Ref. [**11**]. The alcohol composition in the distilling flask starts off at its initial value and approaches zero as the batch distillation proceeds (bottom curves in the figures). The last drop in the bottoms is pure water, although for safety reasons one would not actually want to distill the mixture to dryness. Composition of the distillate is listed in the last three columns of the tables. Three versions of distillate collection are shown. In Column 5, data are recorded at every 5 mL as the distillate is allowed to accumulate in the receiving vessel (top curves). In Column 6, the receiving vessel is changed as soon as each 5 mL of distillate is collected, and the individual distillate samples are kept separate for "analysis" (not plotted) Column 7 represents the instantaneous composition at each temperature and distillate volume noted (middle curves). For the first drop of distillate (0 mL accumulated), all three columns are exactly the same. Composition is also the same, but at a different value, in Columns 5 and 6 for the first 5 mL of accumulated sample. For other entries, compositions show a regular progression from top to bottom and from left to right in Columns 5, 6, and 7.

The atmospheric boiling temperature for a binary mixture depends on the vapor pressures, mol fractions, and activity coefficients of the pure components. Plotting distillate temperature against cumulative distillate from Tables 1 and 2 (Figures 2 and 3) results in the gentle s-shaped curve shown in Fig. 5.8 in Refs. [**6,7**] and Fig. 5.1 of Ref. [**20**]. The simulated experiment has been extended beyond the range of Tables 1 and 2 to collect extra distillate and provide additional temperature data points in the figures. The curves for ethanol-water and methanol-water cross at 100 °C for the last droplet remaining and extrapolate on the other end to the respective boiling points of pure ethanol and pure methanol.

The densities from Figures B1 and B2 in Appendix B can be used to estimate the hot still-pot liquid remaining. Bottoms volume in Column 2 at each boiling temperature is 3-5 % greater than the volume of the same liquid at 20 °C. Change in volume remaining is therefore not accounted for exactly by subtracting the volume of distillate removed, measured at 20 °C.

Table 1. Summary of Simulated Batch Distillation Experiment – Ethanol-Water

Temp (°C)	Bottoms (W) g moles mass (g) mL @ 20 °C *mL @ temp*	Distillate (D) g moles mass (g) mL @ 20 °C	Ethanol in Bottoms (x_W) mol fract wt % *% vol*	Ethanol in Distillate (x_D) Cumulative mol fract wt % *% vol*	Individual mol fract wt % *% vol*	Instantaneous mol fract wt % *% vol*
89.1	9.7367 194.6994 **200.0 *209.5***	0.0000 0.0000 **0**	0.0706 16.269 ***20.0***	0.3704 60.07 ***67.8***	0.3704 60.07 ***67.8***	0.3704 60.07 ***67.8***
89.6	9.5787 190.2358 **195.2 *204.2***	0.1581 4.4636 **5**	0.0658 15.26 ***18.8***	0.3645 59.46 ***67.2***	0.3645 59.46 ***67.2***	0.3577 58.74 ***66.5***
90.0	9.4180 185.7561 **190.3 *199.0***	0.3188 8.9433 **10**	0.0609 14.22 ***17.5***	0.3578 58.76 ***66.5***	0.3513 58.07 ***65.8***	0.3439 57.27 ***65.0***
90.5	9.2542 181.2583 **185.4 *193.7***	0.4825 13.4410 **15**	0.0560 13.17 ***16.3***	0.3508 58.04 ***65.8***	0.3370 56.51 ***64.3***	0.3289 55.63 ***63.5***
91.0	9.0871 176.7406 **180.5 *188.5***	0.6496 17.9588 **20**	0.0511 12.11 ***15.0***	0.3432 57.20 ***65.0***	0.3214 54.78 ***62.6***	0.3128 53.79 ***61.6***
91.6	8.9161 172.2004 **175.6 *183.3***	0.8206 22.4989 **25**	0.0463 11.04 ***13.7***	0.3352 56.32 ***64.1***	0.3045 52.83 ***60.7***	0.2952 51.72 ***59.6***
92.2	8.7409 167.6352 **170.7 *178.0***	0.9959 27.0641 **30**	0.0415 9.96 ***12.3***	0.3265 55.36 ***63.2***	0.2863 50.63 ***58.5***	0.2763 49.40 ***57.2***
92.8	8.5607 163.0420 **165.8 *172.8***	1.1760 31.6573 **35**	0.0367 8.88 ***11.0***	0.3174 54.31 ***62.2***	0.2665 48.17 ***56.0***	0.2559 46.79 ***54.5***
93.5	8.3750 158.4176 **160.8 *167.5***	1.3618 36.2818 **40**	0.0321 7.82 ***9.7***	0.3075 53.10 ***61.0***	0.2454 45.40 ***53.1***	0.2341 43.87 ***51.5***
94.2	8.1830 153.7584 **155.8 *162.3***	1.5538 40.9410 **45**	0.0276 6.77 ***8.3***	0.2971 51.94 ***59.8***	0.2228 42.29 ***49.8***	0.2109 40.61 ***48.0***
94.9	7.9839 149.0609 **150.8 *157.0***	1.7529 45.6385 **50**	0.0233 5.76 ***7.2***	0.2859 50.59 ***58.4***	0.1909 38.84 ***46.1***	0.1867 36.99 ***44.0***

Table 2. Summary of Simulated Batch Distillation Experiment – Methanol-Water

Temp (°C)	**Bottoms (W)** **g moles** **mass (g)** **mL @ 20 °C** ***mL @ temp***	**Distillate (D)** **g moles** **mass (g)** **mL @ 20 °C**	**Methanol in Bottoms (x_W)** **mol fract** wt % *% vol*	**Methanol in Distillate (x_D)**		
				Cumulative **mol fract** wt % *% vol*	**Individual** **mol fract** wt % *% vol*	**Instantaneous** **mol fract** wt % *% vol*
87.9	10.0386 194.8438 **200.0 *206.7***	0.0000 0.0000 **0**	0.0991 16.359 ***20.0***	0.4169 55.97 ***63.8***	0.4169 55.97 ***63.8***	0.4169 55.97 ***63.8***
88.4	9.8485 190.3237 **195.1 *201.7***	0.1902 4.5201 **5**	0.0931 15.26 ***18.7***	0.4100 55.27 ***63.1***	0.4100 55.27 ***63.1***	0.4029 54.54 ***62.3***
88.9	9.6560 185.7881 **190.1 *196.6***	0.3826 9.0558 **10**	0.0871 14.50 ***17.8***	0.4027 54.52 ***62.3***	0.3957 53.79 ***61.6***	0.3880 52.99 ***60.8***
89.5	9.4610 181.2358 **185.2 *191.6***	0.5776 13.6081 **15**	0.0810 13.55 ***16.6***	0.3951 53.74 ***61.5***	0.3807 52.22 ***60.0***	0.3722 51.32 ***59.1***
90.0	9.2634 176.6655 **180.3 *186.5***	0.7753 18.1783 **20**	0.0750 12.60 ***15.5***	0.3872 52.90 ***60.7***	0.3647 50.51 ***58.2***	0.3554 49.51 ***57.2***
90.6	9.0628 172.0760 **175.3 *181.5***	0.9758 22.7678 **25**	0.0690 11.64 ***14.3***	0.3788 52.02 ***59.8***	0.3478 48.67 ***56.3***	0.3376 47.54 ***55.1***
91.2	8.8590 167.4658 **170.4 *176.4***	1.1796 27.3780 **30**	0.0630 10.68 ***13.2***	0.3701 51.09 ***58.8***	0.3299 46.68 ***54.2***	0.3188 45.42 ***52.9***
91.8	8.6517 162.8333 **165.4 *171.4***	1.3869 32.0105 **35**	0.0571 9.72 ***12.0***	0.3610 50.11 ***57.8***	0.3112 44.55 ***52.0***	0.2989 43.12 ***50.5***
92.4	8.4407 158.1771 **160.5 *166.3***	1.5979 36.6668 **40**	0.0513 8.78 ***10.9***	0.3514 49.06 ***56.7***	0.2920 42.31 ***49.6***	0.2780 40.64 ***47.8***
93.1	8.2256 153.4954 **155.5 *161.2***	1.8130 41.3484 **45**	0.0457 7.85 ***9.7***	0.3414 47,96 ***55.6***	0.2721 39.93 ***47.0***	0.2562 37.98 ***44.9***
93.8	8.0061 148.7868 **150.5 *156.1***	2.0325 46.0570 **50**	0.0402 6.94 ***8.6***	0.3310 46.80 ***54.4***	0.2517 37.43 ***44.3***	0.2335 35.14 ***41.7***

Density of the residual liquid (ρ BOTTOMS) at each temperature and composition is plotted against temperature in Figure 5. The curves intersect at the liquid density of pure water boiling at 100 °C and 1 atm. "Experimental" points extrapolate smoothly to this datum along the curves drawn in the figure. At the other end, the curves extrapolate to the densities of pure methanol and pure ethanol at their respective atmospheric boiling points.

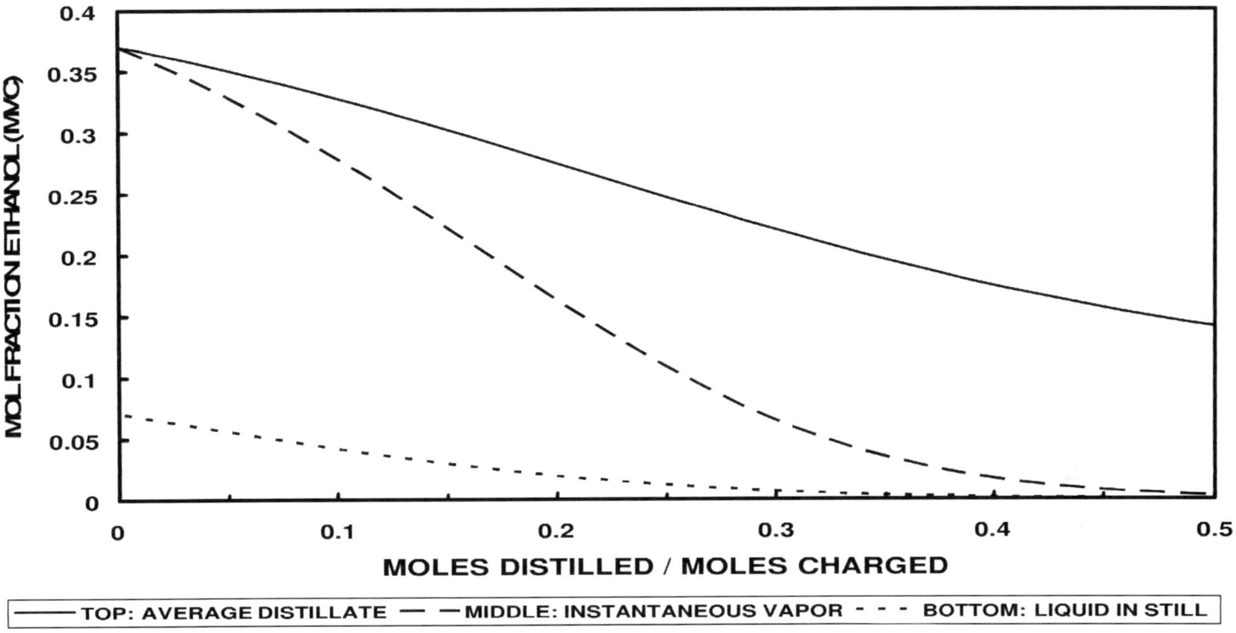

Figure 2. Batch Distillation of a 20 Volume % Ethanol-Water Solution at 1 Atm Pressure

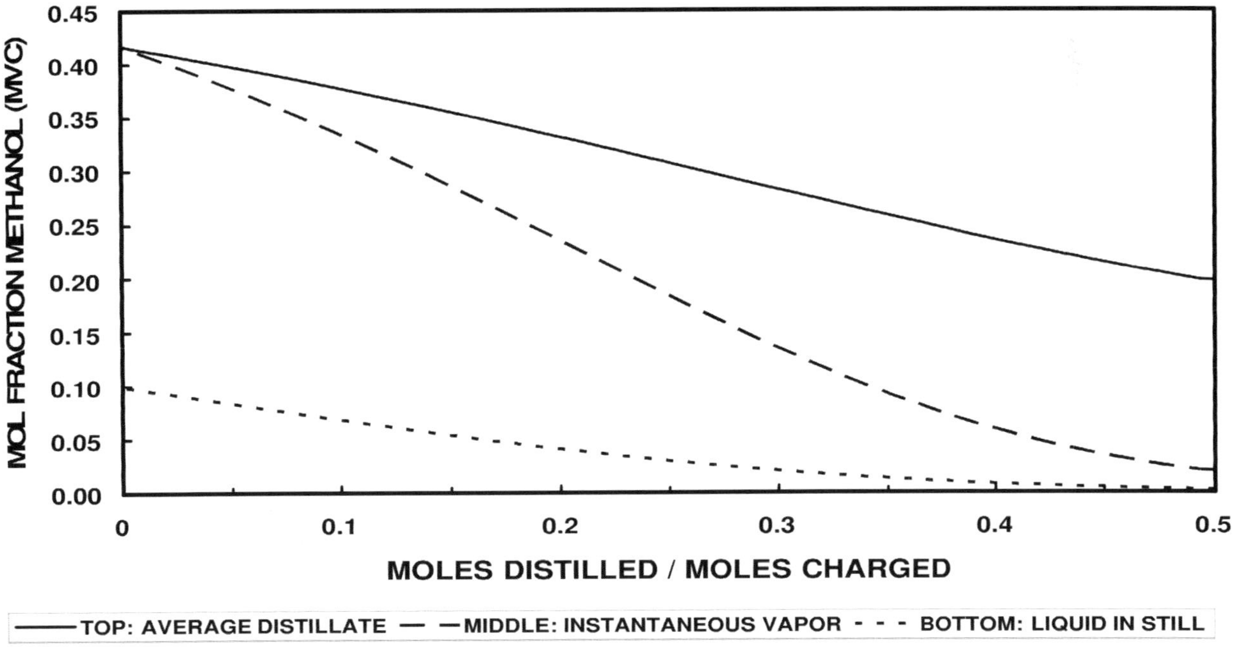

Figure 3. Batch Distillation of a 20 Volume % Methanol-Water Solution at 1 Atm Pressure

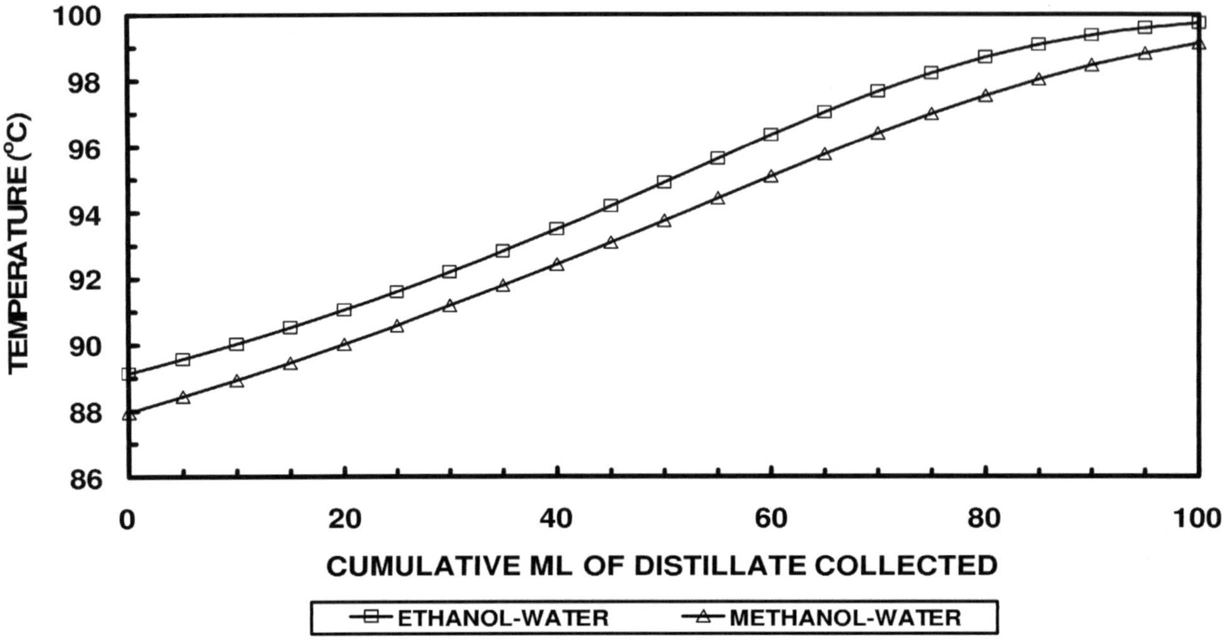

Figure 4. Still-Pot Temperature vs. Cumulative Volume of Distillate for Simulated Ethanol-Water and Methanol-Water Batch Distillations at 1 Atm Pressure

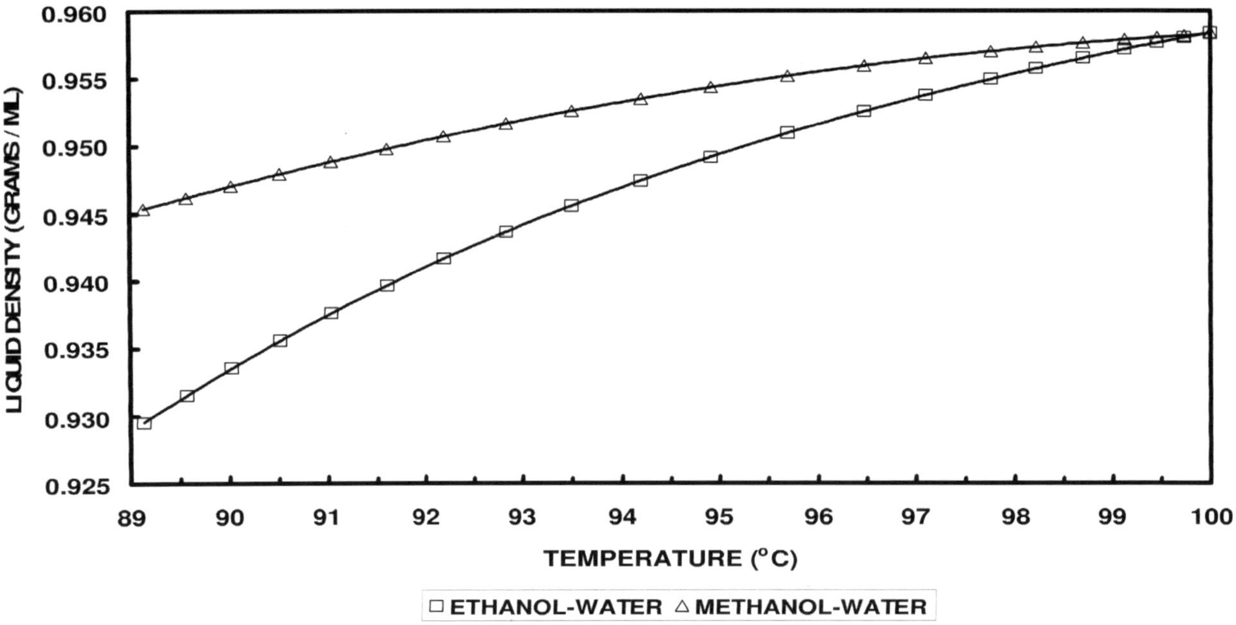

Figure 5. Estimated Density of Solution Remaining in the Still Pot at Boiling Temperature for Simulated Methanol-Water and Ethanol-Water Batch Distillations at 1 Atm Pressure

The empirical curve-fit function for ethanol-water, forced to pass through the 100 °C point:

$$\rho_{\text{BOTTOMS}} = 0.958380 + 1.35444 \times 10^{-3}\,(t - 100\ °C) - 8.1550 \times 10^{-5}\,[(t - 100\ °C)]^{2} - 3.23156 \times 10^{-7}\,[(t - 100\ °C)]^{4} \qquad (17)$$

reproduces the points of the simulation with a correlation coefficient (r^2) of 100 % and a standard deviation (s^2) of 0.00002. For methanol-water, the function

$$\rho_{\text{BOTTOMS}} = 0.0201463 + 0.0147232 \times (t\ °C) - 0.00007155 \times (t\ °C)^{2} \qquad (18)$$

reproduces the points of the simulation with a correlation coefficient (r^2) of 100 % and a standard deviation (s^2) of 0.00008370. Calculated ρ_{WATER} at 100 °C is 0.95828 vs. 0.95838. These functions allow one to estimate the bottoms density (and therefore volume remaining) from temperature alone when batch-distilling without reflux a 20 vol % (60 °F, 15.56 °C) of ethanol-water or methanol-water solution. Division of mass by density gives volume.

In the laboratory, distillate solutions cooled to a standard temperature of perhaps 20 °C would be analyzed quantitatively by any of the methods discussed in Appendix B. For example, densities at laboratory temperatures are well known. Any samples extracted from the distilling flask should also be cooled to laboratory temperature before analysis by density.

On a qualitative basis, unheated ethanol-water distillate samples of about 50 % by volume or more would light off at laboratory temperature with a small flame and leave no residue on the watch glass (Appendix B). This qualitative ignition test should produce positive results at every 5 mL increment for the cumulative samples and for most of the individual samples (Table 1).

In some versions of the experiment, the composite ethanol-rich distillate is saved and further purified using a separate refluxed fractionating column. In a properly designed column, composition will approach pure water in the bottoms and the azeotrope in the overhead. A sample of that distillate would ignite on the watch glass as well, burning more vigorously with a brighter flame and again leaving no visible residue.

Based on flash-point data, anecdotal evidence from the Internet, and results of outdoor flammability tests (all discussed in Appendix B), all of the calculated methanol-water distillate compositions in Table 2 would likely be flammable when the ambient temperature is about 30 °C (86 °F) or above. In addition, the cumulative samples and most of the individual samples would be expected to ignite at a lower ambient temperature in the laboratory. Methanol fires are particularly dangerous because the flame is invisible in daylight to the naked eye **[21]**; combustible paper has been used to reveal the invisible flame **[17]**.

Wrap-up

The newly developed equation has successfully facilitated the simulation of unrefluxed batch distillation experiments; however, modeling of any further purification of distillate samples in a fractionating column is beyond the scope of the present investigation.

Summary and Conclusions

- Batch, or differential, distillation without reflux is described by the Rayleigh Equation:
  - + Relates composition and amount of material remaining in distilling flask
  - + Other quantities determined by material balance.
- Numerical integration is required for the Rayleigh Equation in its basic form.
- Substitution of relative volatility (α) allows analytical integration for constant α.
- New equation derived here allows analytical integration whether α is constant or not.
- The new function was used in simulation of batch distillation of two alcohols and water.
- Ethanol-water and methanol-water results compare favorably with expectations, e.g.,
  - + Boiling point vs. composition curve
  - + Still pot temperature vs. cumulative volume of distillate collected
  - + Density of solution remaining in still pot.
- Methods to analyze liquid composition are discussed for both systems:
  - + Gas chromatography (GC).
  - + Refractive index data compiled from various sources.
  - + Densities of alcohol-water solutions extended beyond range of published data.
  - + Qualitative ethanol-water and methanol-water ignition test / flash points.

About The Author

Robert G. Kunz was an environmental engineering manager at Air Products and Chemicals, Inc., Allentown, PA before retiring after 26+ years of service. He then joined Cormetech, Inc., Durham, NC as Technical Project Manager in support of sales and marketing efforts in the petroleum refining and petrochemical industries. He held engineering positions previously at Esso Research and Engineering Company, Florham Park, NJ and The M.W. Kellogg Company, New York, NY and is currently an independent environmental consultant. "Dr. Bob" has earned a BChE degree in Chemical Engineering from Manhattan College, a PhD in Chemical Engineering from Rensselaer Polytechnic Institute, an MS in Environmental Engineering from Newark College of Engineering, and an MBA from Temple University. He has contributed numerous publications to the technical literature, including topics in phase equilibria, and is a recipient of the Water Pollution Control Federation's Harrison Prescott Eddy Medal for noteworthy research in wastewater treatment. He is author or co-author of two books and holder of one U.S. patent. He is a member of the American Institute of Chemical Engineers (AIChE), the American Chemical Society (ACS), and the Air & Waste Management Association (A&WMA) and is a licensed professional engineer in several states.

Acknowledgement

The author is grateful to Roberta Kunz Fox, AIA, of fox^2 design for help with Figure 1.

References

1. Kunz, R.G., "The Rayleigh Equation Revisited: What to Do When Alpha (α) Isn't Constant," Paper 118b presented at the 2012 A.I.Ch.E. Spring Meeting, Houston, TX (April 1 - 5, 2012).

2. Kroschwitz, J.I. and M. Howe-Grant, editors, "Kirk-Othmer Encyclopedia of Chemical Technology," 4 ed., Vol. 8, pp.311-322, Wiley, New York (1993).

3. Coulson, J.M. and J.F. Richardson with J.R. Backhurst and J.H. Harker, "Chemical Engineering," 4 ed., Vol. 2, pp.436-440, Butterworth-Heinemann, Oxford (1999).

4. Zubrick, J.W., "The Organic Chem Lab Survival Manual," 7 ed., 350 pp., Wiley, Hoboken, NJ (2008).

5. Mayo, D.W., R.M. Pike, and D.C. Forbes, "Microscale Organic Laboratory with Multistep and Multiscale Syntheses," 5 ed., p.61, Wiley, Hoboken, NJ (2011).

6. Williamson, K.L., R. Minard, and K.M. Peters, "Macroscale and Microscale Organic Experiments," 5 ed., pp.88-104, Houghton Mifflin Company, Boston, MA (2007).

7. Williamson, K.L., R. Minard, and K.M. Peters, "Macroscale and Microscale Organic Experiments," Selected Material Special Edition for University of Massachusetts, Amherst, 5 ed., pp.88-104, Houghton Mifflin Company, Boston, MA (2007).

8. Lord Rayleigh, *Phil. Mag.*, "On the Distillation of Binary Mixtures," Series 6, **4(23)**, 521-537 (Nov. 1902).

9. Perry, J.H., editor, "Chemical Engineers' Handbook," 3 ed., 1942 pp., McGraw-Hill, New York (1950).

10. Badger, W.L. and J.T. Banchero, "Introduction to Chemical Engineering," pp. 245-258, McGraw-Hill, New York (1955).

11. McCabe, W.L. and J.C. Smith, "Unit Operations of Chemical Engineering," 3 ed., pp. 484-487, 535-540, 606, McGraw-Hill, New York (1976).

12. Thomas, G.B., Jr., "Calculus and Analytic Geometry," 3 ed., pp.385-387, Addison-Wesley Publishing Company, Inc., Reading, MA (1960).

13. Shoemaker, D.P., C.W. Garland, and J.W. Nibler, "Experiments in Physical Chemistry," 5 ed., 890 pp., McGraw-Hill, New York (1989).

14. Hodgman, C.D., editor-in chief, "Mathematical Tables from Handbook of Chemistry and Physics," p.249, Chemical Rubber Publishing Co., Cleveland, OH (1958).

15. Fleser, L.F. and K.L. Williamson, "Organic Experiments," 7 ed, pp.64-68, D.C. Heath and Company, Lexington, MA (1992).

16. Anon., "Ethanol," www.worldlingo.com/ma/enwiki/en/Ethanol (accessed on 3/8/2011).

17. "Water Meth is it Safe question asked.. Flammability etc." – SEAT Cupra.net – SEAT Forum www.seatcupra.net/forums/showthread.php?t=360226 (accessed on 5/8/2014).

18. Flick, E.W., editor, "Industrial Solvents Handbook," 5 ed., 994 pp., Noyes Data Corporation, Westwood, NJ (1998).

19. Green, D.W. and J.O. Maloney, "Perry's Chemical Engineer's Handbook,"6 ed,.p.3-88, McGraw-Hill, New York (1984).

20. "5. Distillation and Boiling Points," http://courses. chem. psu/chem36/ Experiments/PDF's_ for_ techniques/Distillation_&_BP.pdf (accessed on 4/5/2014).

21. "Water Meth is it Safe question asked.. Flammability etc." – SEAT Cupra.net – SEAT Forum www.seatcupra.net/forums/showthread.php?t=4015096 (accessed on 5/8/2014).

22. Dean, J.A., editor, "Lange's Handbook of Chemistry," 11 ed., McGraw-Hill, New York (1973).

23. Weast, R. C. and M. J. Astle, "CRC Handbook of Chemistry and Physics," 62 ed., CRC Press, Inc., Boca Raton, FL (1981).

24. Dean, J.A., editor, "Lange's Handbook of Chemistry," 12 ed., McGraw-Hill, New York (1978).

25. Dean, J.A., editor, "Lange's Handbook of Chemistry," 13 ed., p.4-59 (water) and Table 7-15 (organics), McGraw-Hill, New York (1985).

26. Kunz, R.G., "Environmental Calculations A Multimedia Approach," pp.41-52, Wiley, Hoboken, NJ (2009).

27. Carlson, H.C. and A.P. Colburn, "Vapor-Liquid Equilibrium of Nonideal Solutions – Utilization of Theoretical Methods to Extend Data," *Ind. Eng. Chem.*, **34(5)**, 581-589 (May 1942).

28. Beebe, A.H., Jr., K.E. Coulter, R.A. Lindsay, and E.M. Baker, "Equilibria in Ethanol-Water System at Pressures Less Than Atmospheric," *Ind. Eng. Chem.*, **34(12)**, 1501-1504 (Dec. 1942).

29. Otsuki, H. and F.C. Williams, "Effect of Pressure on Vapor-Liquid Equilibria for the System Ethyl Alcohol-Water," *CEP Symposium Series No. 6*, **49**, 55-67 (1953).

30. Othmer, D.F., W.P. Moeller, S.W. Englund, and R.G. Christopher, "Composition of Vapors from Boiling Binary Solutions," *Ind. Eng. Chem.*, **43(3)**, 707-711 (Mar. 1951).

31. Speight, J.G., editor, "Lange's Handbook of Chemistry," 16 ed., p.1.95 and p.2.294, McGraw-Hill, New York (2005).

32. Department of Commerce, Bureau of Standards, "Standard Density and Volumetric Tables – Circular No. 19," 6 ed., 72 pp., U.S. Government Printing Office, Washington, DC (Oct.31, 1924).

33. Washburn, E.W., editor-in-chief, "International Critical Tables of Numerical Data, Physics, Chemistry and Technology," 1 ed., Vol. III, 444 pp., McGraw-Hill, New York (1928).

34. Carr, C. and J.A. Riddick, "Physical Properties of Methanol-Water System," *Ind. Eng. Chem.*, **43**(**3**), 692-696 (Mar. 1951).

35. Mikhail, S.Z. and W.R. Kimel, "Densities and Viscosities of Methanol-Water Mixtures," *J. Chem. Eng. Data*, **6**(**4**), 533-537 (Oct. 1961).

36. Sakuri, M. and T. Nakagawa, "Densities of Dilute Solutions of Water in Benzene and in Methanol at 278.15, 288.15, 298.15, 308.15, and 318.15 K. Partial Molar Volumes and Vw and Values of ∂Vw/∂T for Water in Benzene and in Methanol," *J. Chem. Thermodynamics*, 14, 269-274 (1982).

37. Shoemaker, D.P., C.W. Garland, and J.W. Nibler, "Experiments in Physical Chemistry," 5 ed., 890 pp., McGraw-Hill, New York (1989).

38. "The New Encyclopædia Britannica," Vol. 13, p.525, Encyclopædia, Inc., Chicago, IL (2002).

39. Kenkel, J., "Analytical Chemistry for Technicians," 3 ed., 554 pp., Lewis Publishers, Boca Raton, FL (2003).

40. Jones, C.A. E.M. Schoenborn, and A.P. Colburn, "Equilibrium Still for Miscible Liquids," *Ind. Eng. Chem.*, **35**(**6**), 666-672 (June 1943).

41. Schweppe, J.L. and J.R. Lorah, "Ternary System Ethyl Alcohol – *n*-Heptane – Water at 30 °C," *Ind. Eng. Chem.*, **46**(**11**), 2391-2392 (Nov. 1954).

42. Gray, D.E., editor, "American Institute of Physics Handbook, pp.2-141,2-142,2-152,2-153,2-157, McGraw-Hill, New York (1957).

43. "Ethyl Alcohol Handbook, 6 ed., Lyondell Chemical Company, Houston. TX, www.itecref.com/pdf/**Ethyl_Alcohol**_Handbook_Equistar.pdf (accessed on 3/8/2011).

44. Brochure "Technical Information & Safe Handling Guide for Methanol," Version 3.0, Methanex Corporation, Vancouver, B.C., Addison, TX (Sept. 2006).

45. Palleros, D.R., "Experimental Organic Chemistry," 833 pp., Wiley, New York (2000).

46. Dean, J.A., editor, "Analytical Chemistry Handbook," p.9.1, McGraw-Hill, New York (1995).

47. J. Ortega, J., "Densities and Refractive Indices of Pure Alcohols as a Function of Temperature," *J, Chem. Eng. Data*, **27**(**3**), 312-317(July 1982).

48. Andrews, L.W., "The Refractive Indices of Alcohol-Water Mixtures," *J. Am. Chem. Soc.*, **30**(**3**), 353-360 (1908).

49. Nowakowsha, J., "The Refractive Indices of Ethyl Alcohol and Water Mixtures," Masters Theses Loyola University Chicago, IL (1939). Paper 668. http://ecommons.luc.edu/luc_thesis/668.

50. "Distillation Course CHEE 4500 (Spring 2004)" https://www.myedu.com/UT-The-University-of-Toledo/CHEE-4500-Chemical-Engineering-Laboratory-I/course/ 290063/ (accessed 4/13/2014).

51. Balke, C., N. Butler, and K. Courtright, Distillation Column and Refractometer Calibration," Oklahoma State University CHE 4002 Project I, 26 pp. (Feb. 10, 2010). (accessed 4/16/2014).

52. Ho, V. and M. Butzow, "Methanol Recovery via Distillation -- Rotation 1," 36 pp. (Oct. 6, 2005). http://chem.engr.utc.edu/ench435/2005/CHBE-430-Distillation-Final-Report.htm. (4/17/2014).

53. Doroschewsky and Dvorsehantschik, *J. Russ. Phys. Chem. Soc.*, **40**, 908-931 (1908); *Chem. Zentr.*, **2**, 1569-1571 (1908), as cited in Simmonds, C., "Alcohol, Its Production, Properties, Chemistry, and Industrial Applications," Macmillan, London (1919).

54. Speight, J.G., editor, "Lange's Handbook of Chemistry," 16 ed., p.1.95 and p.2.294, McGraw-Hill, New York (2005).

55. Washburn, E.W., editor-in-chief, "International Critical Tables of Numerical Data, Physics, Chemistry and Technology," 1 ed., Vol. VII, 506 pp., McGraw-Hill, New York (1930).

56. Openhaim, G. and E. Grushka, "Temperature-dependent Refractive Index Issues Using a UV – visible Detector in High Performance Liquid Chromatography," *J. Chromatology A*, **942**, 63-71 (2002).

57. "Ethyl Alcohol Handbook," 6 ed., Lyondell Chemical Company, Houston. TX, www.itecref.com/pdf/**Ethyl_Alcohol**_Handbook_Equistar.pdf (accessed on 3/8/2011).

58. Anon., "Alcoholic Proof," http://www.bookrags.com/wiki/proof-alcohol (accessed on 3/3/2011).

59. U.S. Census Bureau, Foreign Trade Statistics, "2011 Schedule B – Chapter 22," http://www.census.gov/foreign-trade/schedules/b/2011/c22.html (accessed 3/8/2011).

60. Anon., "At What 'Proof' Will Spirits Burn? – Alcohol Fire Chemistry," http://ask.metafilter.com/95484/At-what-proof-will-spirits-burn (accessed on 3/8 2011).

61. Spencer, A.B. and G.R. Colonna, editors, "Fire Protection Guide to Hazardous Materials," 13 ed., p. 325-57, p. 325-81, p. 497-8, p. 497-10, NFPA, Quincy, MA (2002).

62. "Materials Safety Data Sheet Methanol," Terra Nitrogen Corporation, Sioux City, IA (April 1, 2001).

63. "Materials Safety Data Sheet 10% Methanol / 90% Water Volume / Volume," Honeywell, Burdick & Jackson, Muskegon, MI (Oct. 2005).

64. "Flammability of Methanol-Water Mixtures" – waterinjection.info forum www.aquamist.co.uk/vbulletin/showthread.php?t=856 (accessed on 5/7/2014).

65. "Water Meth is it Safe question asked.. Flammability etc." – SEAT Cupra.net – SEAT Forum www.seatcupra.net/forums/showthread.php?t=4015096 (accessed on 5/8/2014).

66. "Flash Points of Methanol based Water Solutions," http://www.engineeringtoolbox.com/methanol- water-d_987.html (accessed on 4/23/2014).

APPENDIX A

Summary of Vapor-Liquid Equilibrium (VLE) Relationships

Low-pressure binary vapor-liquid equilibrium is described by the following relationships involving temperature (t), total pressure (P_T), partial pressure (p), vapor pressure (p^{vap}), and liquid- and vapor-phase phase mol fractions (x and y) of each component, plus so-called activity coefficients (γ's) when a given system cannot be considered to be ideal. The subscript 1 denotes the more volatile component (MVC); 2 the less volatile component (LVC),

Ideal Systems

A system is said to be ideal when it follows Raoult's Law [Equations (A1) and (A2)]

$$p_1 = p_1^{vap} x_1 \tag{A1}$$

$$p_2 = p_2^{vap} x_2 \tag{A2}$$

and

$$P_T = p_1 + p_2 \tag{A3}$$

with

$$x_1 + x_2 = 1 \tag{A4}$$

If the Ideal Gas Law and Dalton's Law are assumed to apply to the vapor phase, vapor-phase mol fractions (y_1 and y_2) can be calculated from:

$$y_1 = p_1 / P_T = p_1^{vap} x_1 / P_T \tag{A5}$$

$$y_2 = p_2 / P_T = p_2^{vap} x_2 / P_T \tag{A6}$$

with

$$y_1 + y_2 = 1 \tag{A7}$$

Vapor Pressure. Pure-component vapor pressures in both ideal and non-ideal systems are conveniently calculated using the Antoine Equation:

$$\log_{10} p^{vap}\,(\text{mmHg}) = A - B / [t\,(°\text{C}) + C] \tag{A8}$$

Antoine constants A, B, and C are listed in Table A1 for the chemicals under consideration here.

Calculation of x-y Curve. By combining and rearranging the equations above, one obtains:

$$x_1 = (P_T - p_2^{vap}) / (p_1^{vap} - p_2^{vap}) \tag{A9}$$

Chiefly by means of this equation, the entire x-y equilibrium curve and temperature-composition curves at constant pressure for ideal systems can be mapped out in a straightforward manner. One first chooses a temperature, computes vapor pressures from Equation (16), then calculates x_1 from Equation (A9) and y_1 from Equation (A5). Only if a specific value of x_1 is desired, does the calculation become trial and error, in which a new value of t is repeatedly chosen until the desired value of x_1 is obtained to within whatever tolerance is satisfactory.

Table A1. Antoine Equation Constants of Interest [a]

Chemical Name and Synonyms	Formula	MW [b]	BP (°C) [c]	Applicable Range(°C)	A	B	C
Ethanol Ethyl Alcohol Grain Alcohol Methyl Carbinol	C_2H_5OH	46.07	78.3	—	8.04494	1554.3	222.65
Methanol Methyl Alcohol Wood Alcohol Carbinol	CH_3OH	32.04	64.7	–20 to +140	7.87863	1473.11	230.0
Water Hydrogen Oxide	H_2O	18.02	100.00	0 to 60 60 to 150	8.10765 7.96681	1750.286 1668.21	235.0 228.0

Notes:

(a) Antoine Constants from "Lange's Handbook of Chemistry", 11th Edition [**22**, Table 10-10].

(b) Molecular Weight (MW) from "Handbook of Chemistry and Physics," 62nd Edition [**23**, pp. C-65 to C-576], rounded for water [**23**, p.B-105] to 2 decimal places.

(c) Boiling point (BP) at 760 mmHg from "Lange's Handbook," 11th-13th Editions [**22**, p.4-59 and Table 7-4; **24**, p.4-59 and Table 7-4; **25**] or "Handbook of Chemistry and Physics," 62nd Edition [**22**, pp. C-65 to C-576, p.B105], chosen to be the most consistent with boiling point calculated using the Antoine Constants listed.

Methanol-Ethanol. Ideal systems occur with members of homologous series, isomers, and in certain other cases [**9**]. Because of the similarity of the two molecules, interactions between like and unlike species are nearly the same. Hence, each component behaves as if independent of the other. The classic system used to illustrate Raoult's Law is benzene-toluene. It is highlighted in many textbooks including Ref. [**11**]. Likewise, methanol and ethanol, of interest here, form an ideal binary system with each other. Its x-y curve, published previously [**26**], is drawn in Figure A1 along with experimental data from several investigators.

Non-Ideal Systems

For these, a so-called activity coefficient (γ) is introduced into Raoult's Law to yield:

$$p_1 = p_1^{vap} \gamma_1 x_1 \tag{A10}$$

$$p_2 = p_2^{vap} \gamma_2 x_2 \tag{A11}$$

and

$$P_T = p_1 + p_2 \tag{A3}$$

with

$$x_1 + x_2 = 1 \tag{A4}$$

Under the Ideal-Gas and Dalton's Law assumptions, vapor phase mol fractions are given by:

$$y_1 = p_1 / P_T = p_1^{vap} \gamma_1 x_1 / P_T \quad \text{(A12)}$$

$$y_2 = p_2 / P_T = p_2^{vap} \gamma_2 x_2 / P_T \quad \text{(A13)}$$

again with

$$y_1 + y_2 = 1 \quad \text{(A7)}$$

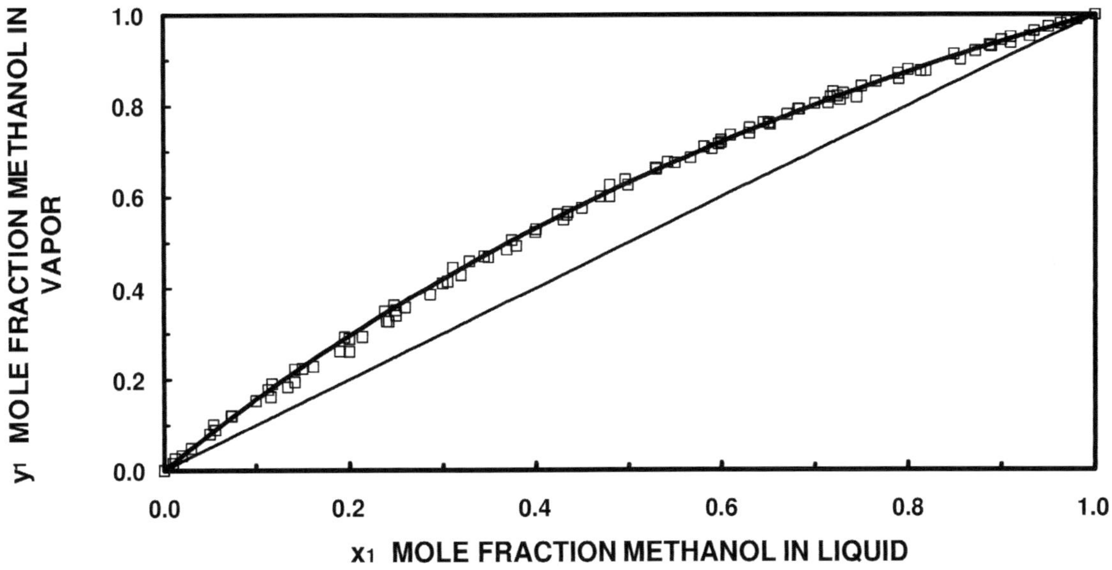

Figure A1. Vapor-Liquid Equilibrium Diagram Methanol-Ethanol at 1 Atm Pressure

Activity Coefficients. Activity coefficients (γ's) account for deviations from ideality. They are back calculated as point values from experimental data and then fitted to a special type of continuous function for use in VLE calculations. There are several empirical activity-coefficient formulations in widespread use; for any given application, some fit the deviations better than others across the range of data. The van Laar equations **[2**; **9**, p.527, **27]**, for example:

$$\log_{10} \gamma_1 = \frac{A_{1\text{-}2}}{[1 + (A_{1\text{-}2}\, x_1) / (A_{2\text{-}1}\, x_2)]^2} \quad \text{(A14)}$$

and

$$\log_{10} \gamma_2 = \frac{A_{2\text{-}1}}{[1 + (A_{2\text{-}1}\, x_2) / (A_{1\text{-}2}\, x_1)]^2} \quad \text{(A15)}$$

fit the data well for the non-ideal systems considered here. These particular equations are also written in terms of natural logarithms (base e), using constants 2.302585... × $A_{1\text{-}2}$ and $A_{2\text{-}1}$.

Methanol-Water. Activity coefficients for methanol-water at 1 atm pressure derived from experimental data are given on p.528 in Perry's 3rd Edition **[9]**. Coefficients $A_{1\text{-}2}$ and $A_{2\text{-}1}$ of are 0.36 and 0.22 ($\log_{10}$ based) or 0.83 and 0.51 (ln based). The previously published **[26]** methanol-water x-y curve with accompanying experimental data from a number of investigators is reproduced in Figure A2. It resembles the curves for ideal systems. It uses the van Laar constants enumerated above and Antoine constants from Table A1, plus a trial-and-error† procedure to compute each point on the continuous curve at 1 atm total pressure.

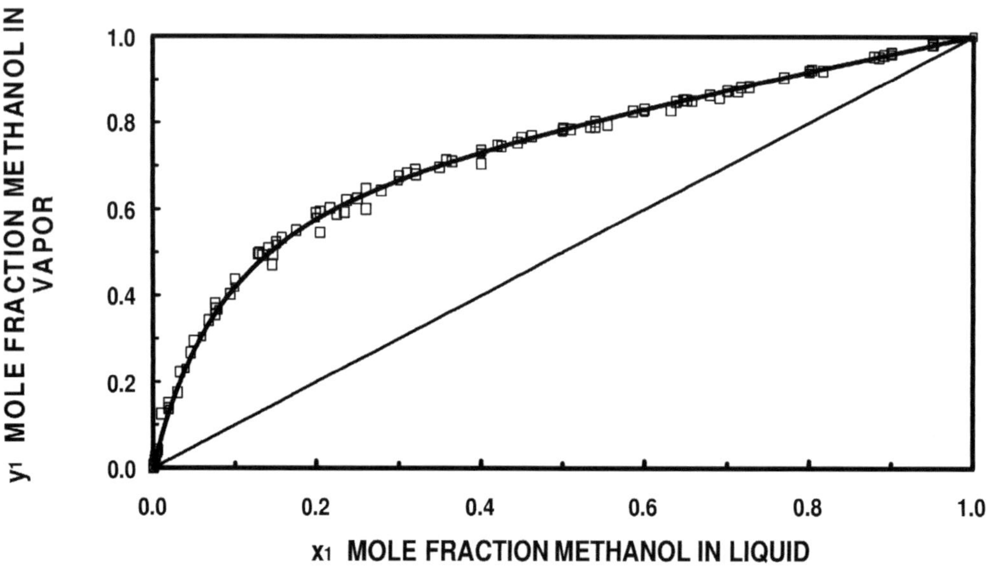

Figure A2. Vapor-Liquid Equilibrium Diagram for Methanol-Water at 1 Atm Pressure

Ethanol-Water. The typical example frequently cited to illustrate an azeotrope, or constant boiling mixture, is the ethanol-water system. This system is well investigated in the literature. Its predicted x-y diagram at atmospheric pressure, shown in Figure A3, was drawn by using the same trial-and-error procedure described above for methanol-water.

It exhibits an azeotrope composed of 95.6 wt % ethanol and 4.4 wt % water at 1 atm pressure (760 mmHg, 1.013 bar) and 78.15 °C; this corresponds to 89.43 mol % ethanol and 10.57 mol % water [**9**, pp.631,633]. This cannot be purified further to 100 % ethanol by distillation at atmospheric pressure. The azeotropic point is different at other total pressure levels [**9**, p.631]; water content decreases with decreasing pressure, and at least in theory becomes zero when distillation pressure is reduced low enough [**9**, p.631;**40**].

In the special case of an azeotrope, one can obtain the van Laar constants for use in Equations (A14) and (A15) from the composition of the azeotropic point **[26]**. For the ethanol-water system, coefficients of 0.68381, 0.41724 ($\log_{10}$ based) or 1.5745, 0.9607 (ln based)

† Constant-pressure calculations for non-ideal systems are always trial and error since the activity coefficients are functions of liquid composition. An efficient method to map out the x-y equilibrium curve is to select a value of x_1 and total pressure and then vary the temperature until the calculated total pressure from Equation (A3) agrees as closely as desired with the total pressure selected.

were determined from the azeotrope at 1 atm in conjunction with Antoine Equation vapor pressures calculated using the ethanol and water constants of Table A1. Since the fitting procedure to obtain them involves only the data at the azeotropic point, agreement with other experimental values along the curve is not guaranteed. However, the myriad of accompanying data points from multiple investigators in Figure A3 are so numerous as to obscure the calculated x-y equilibrium curve almost completely. Data were not selectively chosen to show agreement with the curve, as evidenced by the few stray points on the graph; rather, complete data sets were added one after another until the underlying computer file filled up.

Furthermore, the ethanol-water system at constant pressure has been investigated from below atmospheric pressure **[28]** up to at least 300 lb_f/in^2 absolute (psia) **[29,30]**. It is well described in Ref. **[29]** by van Laar constants only slightly different from those obtained here and differing among themselves but little for huge changes in pressure. Therefore, small day-to-day variations in barometric pressure encountered in the laboratory would alter only P_T in simulation calculations, and not require modification of the 760-mmHg van Laar constants.

Relative Volatility. The relative volatility (α) in Equation (5) can be calculated at constant pressure for any temperature or the corresponding value of composition (x_1) along the VLE curves for ideal systems from a ratio of vapor pressures [Equation (A16)], and for non-ideal systems through the addition of activity coefficients [Equation (7)]:

$$\alpha = (y_1/x_1) \;/\; (y_2/x_2) = p_1^{vap} \;/\; p_2^{vap} \qquad \text{(A16)}$$

$$\alpha = (y_1/x_1) \;/\; (y_2/x_2) = (p_1^{vap}\,\gamma_1) \;/\; (p_2^{vap}\,\gamma_2) \qquad \text{(7)}$$

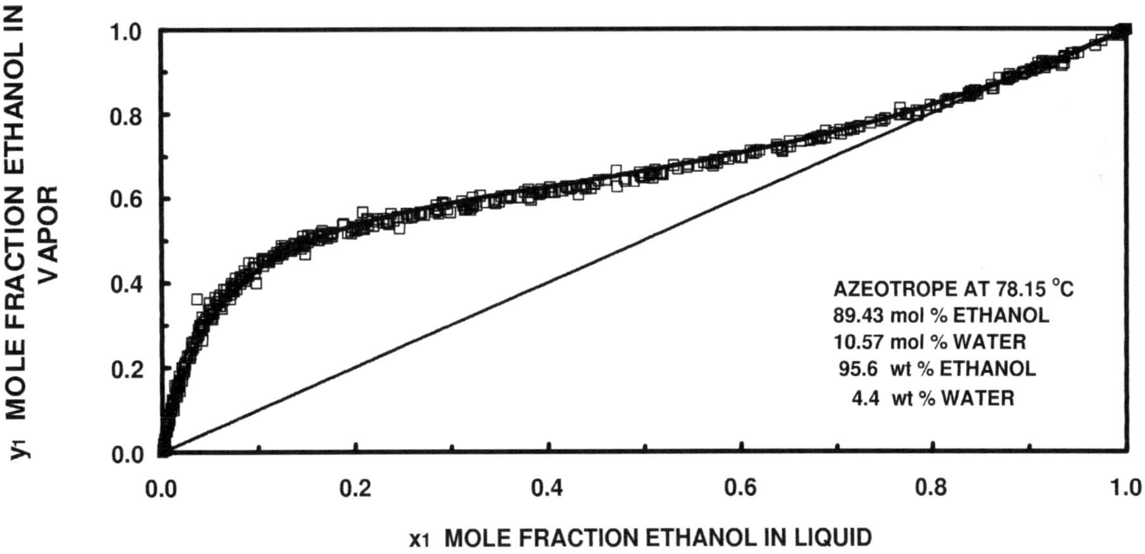

Figure A3. Vapor-Liquid Equilibrium Diagram for Ethanol-Water at 1 Atm

APPENDIX B

ANALYTICAL METHODS FOR ALCOHOL-WATER SOLUTIONS

This appendix reviews several ways to analyze for chemical composition in the ethanol-water and methanol-water systems. These include at least three quantitative methods and one qualitative technique. Quantitative analysis is done by refractive index (RI), density (ρ), or possibly gas chromatography (GC). A qualitative analysis for ethanol is often conducted by attempting to ignite several drops of distillate solution on a watch glass since aqueous solutions of ethanol greater than about 50 vol % are combustible at room temperature **[16]**. Methanol-water combustion characteristics are similar. Topics are discussed in turn below.

Analysis by Density

Published densities of solutions of ethanol and water **[9,22-25,31-33]** and methanol and water **[9,22-25,31-36]** facilitate chemical analysis by careful measurement of density at constant temperature. Unlike the refractive index of alcohol-water solutions discussed below, solution density at a given temperature corresponds to a unique value of composition. The simplest technique employs a pycnometer, or specific gravity bottle, a piece of calibrated glassware of precisely known volume [**37**, pp.191-193]. It is weighed first empty and then full of solution. Density is calculated by dividing solution weight by solution volume. Pycnometer volume is typically about 10 mL **[38]**, but a 5-mL pycnometer is also available [**39**, p.434]. Related studies used a 10-mL **[28,40]** or 5-mL **[41]** pycnometer to analyze by density.

Advantage: Analysis by density has the advantage of using simple laboratory equipment in widespread use. It takes a minimum of time and provides a unique value for ethanol or methanol composition at a constant known temperature. Disadvantage: In some cases, the sample size needed may be greater than the amount of distillate collected over a short period.

Density-Composition Plots. Density data for ethanol-water and methanol-water mixtures across the entire gamut of compositions and sample temperatures are difficult to find. This is not so much of a problem for analysis by density, but rather for converting from mass to volume at any given temperature and vice versa. Density data (g/mL) for liquid water (0-102 °C) and ethanol-water mixtures are tabulated in NBS Circular 19 **[32]** and handbooks reproducing said tables. For pure ethanol, density or its reciprocal, specific volume, is listed in various tables up to 80 °C [**33**, p.41], and tabular entries and correlating equations are provided for density from 0-78 °C **[42]**. Density data in lb_m/gal is plotted either against °C [**18**, p.286] or °F [**43**, p.62] from 5 to 60 °C, parametric in vol % ethanol. Density data for liquid solutions outside this temperature range could not be found in the published literature.

However, the parametric plots [**18**, p.286; **43**, p.62] appear to be well behaved, consistently spaced, and fairly linear at the upper temperatures, suggesting that missing values might be interpolated or extrapolated with some degree of success. To that end, the density plot has been reconstructed here in units of g/mL between 0 and 100 °C, with wt % ethanol as parameter (Figure B1).

The newly developed curves of Figure B1 are based on the data from NBS Circular 19 **[32]**, the equations of Ref. **[42]**, and the aforementioned graph **[18**, p.286**]**. The fit is exact for all values for pure water, for pure ethanol up to 78 °C, and for tabulated mixture data from 10-40 °C **[32]**, and in reasonable agreement with values read with difficulty from the curves of Ref. **[18**, p.286] between 5 and 60 °C. Two different methods produce the same extrapolated pure ethanol curve from 78 to 100 °C.

The top curve (0 % ethanol, 100 % water) contains the actual maximum point in the density for pure water at about 4 °C. Curve fitting predicts a maximum in density for 20 wt % ethanol at about 2 °C; this may or may not be the case. As drawn in Figure B1, the other curves exhibit a continuous decrease in density with increasing temperature.

Estimated densities for pure ethanol beyond 80 °C and for mixtures below 5 °C and above 60 °C are not based on data and should be used with caution, since extrapolated densities are most likely not exact. In addition, although values are shown in Figure B1 for pure ethanol and for some mixture compositions above their atmospheric boiling points, these solutions do not exist as stable liquids at 1 atm pressure. In that case, the predicted densities correspond to whatever slightly elevated pressure is necessary to maintain the solution in the liquid state. This slight difference in pressure should have a negligible effect on density since enormous pressure increases are necessary to effect a significant change in liquid density **[33**, pp.40-42**]**. Finally, while estimated values of density may be useful to calculate volume from mass or vice versa, density values outside the range of data should not be used for analytical purposes to determine solution composition.

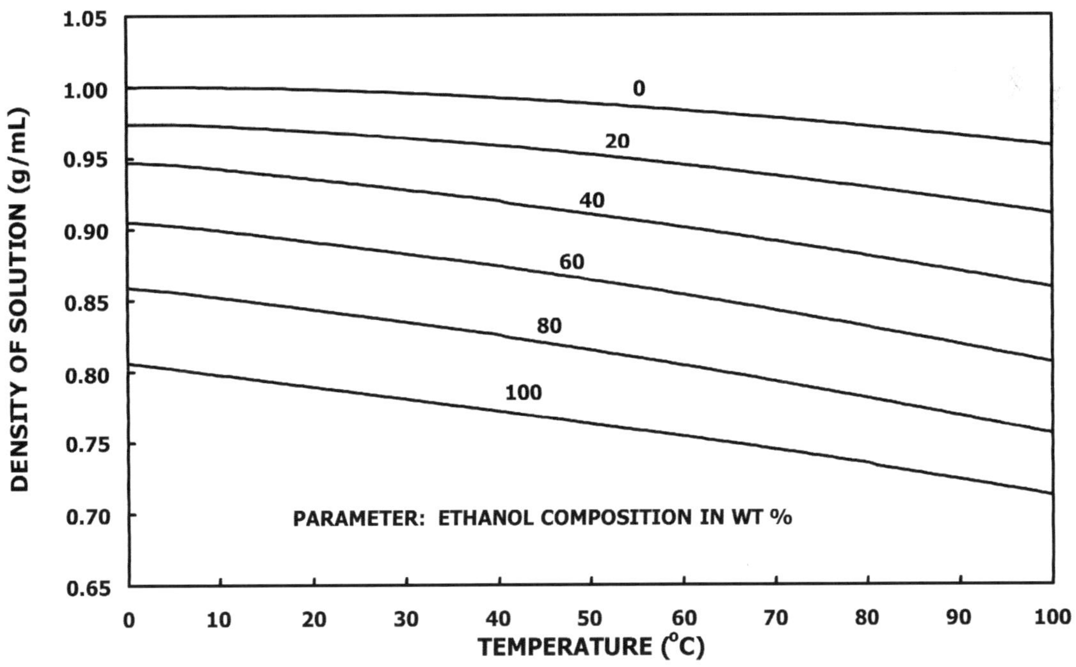

Figure B1. Estimated Liquid Density of Ethanol-Water Solutions

A similar plot of density vs. temperature from +30 °C to –40 °C for methanol-water solutions is available from a methanol manufacturer's website Ref. **[44]**. It too appears to be well behaved and can possibly be extended to the temperature range of interest with the help of other data. Such data exist from 0 °C to 100 °C for pure water as noted above, for 100% methanol **[42,44]** and for methanol-water solutions from 0 °C to 50 °C previously cited. The variation of solution density is said to be linear with mole fraction **[45**, pp.139-140].

Density-composition curves at constant temperature can then be fitted using only pure component densities without needing any data at intermediate compositions; such data where available simply provide a check that the curves are reasonably linear. The slope of a linear density-composition curve varies with temperature and, as it turns out, is well correlated by a quadratic function. The intercept of each density-composition line is the density of pure water at the given temperature, tabulated to 5 decimal places in NBS Circular 19 **[32]**.

The resulting estimated density graph for methanol-water solutions from 0 °C to 100 °C is shown in Figure B2. The same cautions noted above for the ethanol-water curves of Figure B1 apply here also. Solution densities are good only to 2 or perhaps 3 decimal places. Pure component densities, all based on data, are better. Densities for water depicted at the top of the diagram follow the same curve as in Figure B1 (maximum at ~4 °C). Even after all the roundabout calculations, the straight line at the bottom corresponding to 100 % methanol reproduces the starting data, showing a slope of –0.0009+ g/ml per °C, as noted earlier**[34]**.

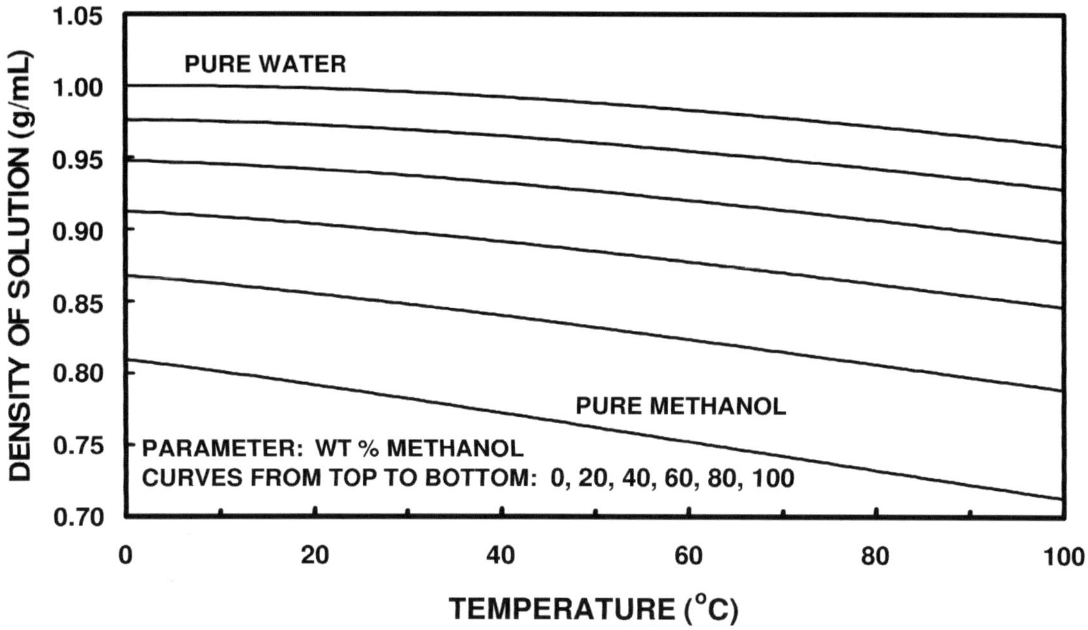

Figure B2. Estimated Liquid Density of Methanol-Water Solutions

Analysis Using Refractive Index

Refractive index (RI) data for mixtures of ethanol and water at various temperatures (n_D^t) are summarized in Table B1, assembled from a number of sources. Refractive index can theoretically be determined to 1 part in 10,000 [**4**, p.319; **6**, p.299] but is more likely to agree with literature values to 1 part in 1,000 because of the presence of impurities [**6**, p.279].

Refractive index is commonly measured at 20 °C (n_D^{20}) [**17**, pp.D-200,D207]. Minor excursions from 20 °C are corrected for by adding a mean value of 0.00045 refractive index units for every °C above 20 °C [**6**,p.299;**46; 47**].

$$n_D^{20} = n_D^t + 0.00045\,(t - 20\ °C) \qquad (B1)$$

As seen in Table B1, differences in refractive index for more appreciable differences in temperature are not exactly accounted for by the general factor.

Refractive index for ethanol-water solutions passes through a maximum point when plotted against composition. This occurs at 79.3 wt % ethanol and 25 °C, as reported by Andrews [**48**], and in the vicinity of 79-80 wt % ethanol at 60 °F (15.56 °C), as determined by curve-fitting the data in Ref. [**24**, p.250] combined with differential calculus. Another study of the maxima is also available [**49**]. A less rigorous inspection of the other entries in Table B1 detects maxima occurring somewhere in a wider range between 70-80 wt % ethanol.

Refractive index measurements for methanol-water solutions are shown in Table B2. Note that some entries in this table relate to volume % rather than to weight %.

$$\text{vol \% } CH_3OH = \text{wt \% } CH_3OH \times \rho_{\text{solution}} / \rho_{\text{pure methanol}} \text{ [34]} \qquad (B2)$$

$$\text{wt \% } CH_3OH = \text{vol \% } CH_3OH \times \rho_{\text{pure methanol}} / \rho_{\text{solution}} \qquad (B3)$$

Density (ρ) of solution and of pure methanol is at the given temperature, and in addition solution density is at the given concentration. Vol % CH_3OH is 100 × volume of pure methanol divided by volume of resulting solution, not by total volume of methanol plus water being mixed.

Maximum points in RI also occur for methanol-water solutions (Table B2). These maxima are also a subject of discussion by others [**50-52**].Refractive indices for methanol-water solutions have been fitted as a function of temperature using a 4th or 5th degree polynomial [**51-52**.]

In the affected range, therefore, more than one alcohol composition corresponds to the same refractive index when the axes of either curve are switched to make refractive index the independent variable for calibration. However, location on the curve can be resolved by adding alcohol or water to the sample and noting whether the refractive index goes up or down.

Advantage: Determination of refractive index requires only two or three drops of sample [**6**, p.300]. Once the procedure is set up and the proper technique is developed [**4**, pp.247-251], it is possible to analyze a new sample every minute [**13**, p.237]. RI should be measured as soon as possible after sampling lest the samples decompose on standing [**6**, p.300]. Disadvantage: Analysis by RI does, however, require specialized equipment [**6**, p.300].

Table B1. Refractive Index (n_D) Values of Ethanol-Water Solutions at Various Temperatures (°C)

Ethanol (wt %)	n_D^{15}	$n_D^{15.56}$	n_D^{20}	n_D^{25}	n_D^{30}	n_D^{40}	n_D^{50}	n_D^{55}
0	1.33345	1.33336	1.33300	1.33250	1.3318	1.3306	1.3290	1.3281
10	1.34020	1.3402	1.3395	1.3389	1.3384	1.3368	1.3349	1.3339
20	1.34778	1.3479	1.3469	1.3462	1.3450	1.3429	1.3406	1.3393
30	1.35470	1.3542	1.3535	1.3520	1.3510	1.3481	1.3452	1.3435
40	1.35948	1.3590	1.3583	1.3565	1.3550	1.3518	1.3484	1.3468
50	1.36290	1.3626	1.3616	1.3598	1.3578	1.3543	1.3506	1.3488
60	1.36505	1.3650	1.3638	1.3620	1.3597	1.3560	1.3522	1.3501
70	1.36645	1.3662	1.3652	1.363038	1.3608	1.3570	1.3528	1.3505
80	1.36690	1.3665	1.3658	1.36331	1.3611	1.3569	1.3525	1.3502
90	1.36626	1.3659	1.3650	1.36239	1.3603	1.3561	1.3515	1.3491
100	1.36332	1.36316	1.3614	1.35941	1.3573	1.3531	1.3487	1.3465

Notes:

(a) Values at 15°C, 30°C, 40°C, 50°C, and 55°C from Ref. [**53**].

(b) Values at 15.56°C (60°F) for pure water and ethanol from Ref. [**18**, p.250]. Other entries at 15.56°C from regression of Ref. [**18**, p.250] values at other compositions.

(c) Values at 20°C from Ref. [**23**, p.D-200 (water) and p.D207].

(d) Values at 25°C either reported by Andrews [**48**], assembled from entries on p.1.95, p.2.294, and Tables 10-71 and 10-72 of Lange's Handbook [**54**], or curve-fitted to interpolate between those values.

Table B2. Refractive Index (n_D) Values of Methanol-Water Solutions at Various Temperatures (°C)

Methanol (wt %)	n_D^{15}	n_D^{20}	Methanol (vol %)	n_D^{25}	n_D^{30}	n_D^{35}	n_D^{40}	n_D^{45}	n_D^{50}
0	1.33339	1.3330	0	1.3325	1.3321	1.3316	1.3309	1.3300	1.3292
10	1.33584	1.3354	10	1.3341	1.3336	1.3330	1.3321	1.3310	1.3303
20	1.33879	1.3381	20	1.3360	1.3357	1.3352	1.3339	1.3330	1.3318
30	1.34138	1.3407	30	1.3382	1.3378	1.3366	1.3352	1.3343	1.3337
40	1.34308	1.3425	40	1.3402	1.3393	1.3376	1.3369	1.3358	1.3342
50	1.34378	1.3431	50	1.3412	1.3400	1.3386	1.3374	1.3360	1.3348
60	1.34327	1.3426	60	1.3412	1.3400	1.3388	1.3372	1.3355	1.3342
70	1.34179	1.3411	70	1.3401	1.3389	1.3371	1.3359	1.3340	1.3318
80	1.33925	1.3385	80	1.3375	1.3356	1.3344	1.3330	1.3313	1.3298
90	1.33545	1.3348	90	1.3333	1.3316	1.3305	1.3290	1.3259	1.3244
100	1.33057	1.3290	100	1.3269	1.3248	1.3235	1.3212	1.3192	1.3168

Notes:

(a) Values at 15°C from Ref. [**55**, p.66]. ICT, Vol. 7, p.66 (1930).
The original reference includes additional entries at intermediate values of ***wt %*** methanol.

(b) Values at 20°C from Ref. [**23**, p.D-200 (water) and p.D217]. HCP, 62 ed., pp.D-200, D-217
The original reference includes additional entries at intermediate values of ***wt %*** methanol.

(c) Data entries between 25°C and 50°C corresponding to values of ***vol %*** methanol from Ref. [**56**].

(d) .All of the above data pass through a maximum point as a function of % methanol.

Analysis by Gas Chromatography (GC)

Gas chromatography is a well-documented separation and analysis technique used in a chemical laboratory. When a liquid sample is injected into the gas chromatograph instrument, it is picked up by a carrier gas, usually helium but possibly nitrogen, and passed through a heated column packed with adsorbent material, where different chemicals are adsorbed and desorbed at different rates. The chemicals being analyzed for arrive at a detector downstream as a series of peaks traced on a strip chart recorder. The time for the arrival of a peak is indicative of each particular chemical in the mixture. The width of each peak, or more properly the area under the peak, represents the amount of that chemical present. More information to supplement this highly simplified explanation is contained in the cited references [**4**, pp.255-274; **6**, pp.215-223; **7**, pp.165-173; **13**, pp.149-151; **45**, pp.144-169].

Advantages: A gas chromatograph can detect extremely low concentrations in an unknown sample. It requires a sample size on the order of 1-25 micro liters (μL), smaller or much smaller than the several drops (20 drops = 1 mL) needed for a refractive index determination.

Disadvantages: The GC instrument is exceptionally expensive to purchase and may have to be shared with other investigators, if it is available at all. It needs an hour or so warm-up time, and requires experience it setting its temperature controls for the particular system being analyzed and some technique/practice in handling the micro-liter syringe to pierce the septum at the instrument's injection port. Residence time for each sample to pass through the column and show up as peaks at the detector may be several minutes or longer.

Summary of Analytical-Quantitative

Properties of alcohol-water solutions from the literature are provided here for reference in conducting laboratory analyses of unknown solution compositions encountered during a distillation experiment. Nevertheless, it is advisable to prepare one's own calibration curves employing the same reagents and laboratory equipment being used to analyze the unknowns. A procedure for preparing standard methanol-water solutions is outlined in Ref. [**39**, p. 458].

Analytical-Qualitative

A qualitative analysis for ethanol in the distillate is often conducted by attempting to ignite several drops of solution on a watch glass. Like hydrogen, ethanol burns with a bluish flame not easily visible in ordinary daylight [**57**, p.101]. Ethanol and its aqueous solutions greater than about 50 % by volume are combustible at room temperature [**16**]. These will ignite and burn to CO_2 and water without leaving any residue.

Alcohol Proof. In former times, this phenomenon formed the basis of a test of the alcoholic spirits given to members of the British Royal Navy as part of their rations [**58**]. In that test, a sample lighting off with gunpowder constituted "proof" that the alcohol had not been watered down. In the U.S. today, *proof* is defined as exactly twice the alcoholic content by volume, measured at 60 °F (15.56 °C) [**18**, Table 6.36; **57**, p. 108; **59**].

Anecdotal evidence suggests that a certain brand of vodka at 100 proof (50 % ethanol by volume) lights off, whereas this same vodka at 80 proof (40 % ethanol by volume) does not **[60]**. That this should be so is explained by the flash point of ethanol-water solutions (Table B3, prepared from a single source **[61]**); flash point is plotted vs. temperature in Ref. **[18**, p.254]. The flash point is the lowest temperature at which just enough liquid is evaporated at the surface to create a combustible concentration in air of gas that will catch fire if an ignition source is present.‡ The table shows that an ethanol-water solution is combustible at "room temperature" at about the stated 50 % by volume. Heating the solution first, which generates additional combustible ethanol vapor (and thereby lowers the flash point), allows solutions of lesser ethanol content to ignite at the same room temperature **[60]**.

Table B3. Flash Points of Ethanol-Water Solutions **[61]**

Ethanol Volume %	Temperature		Ethanol Volume %	Temperature	
	°F	°C		°F	°C
5	144	62	60	72	22
10	120	49	70	70	21
20	97	36	80	68	20
30	85	29	95	63	17
40	79	26	96	62	17
50	75	24	100	55	13

Although not specifically mentioned as an analytical test of distillate composition, some methanol-water mixtures are also combustible, burning with a flame that may not be visible to the naked eye in daylight **[62]**. Methanol solutions containing up to 74 % water are classified by the National Fire Protection Association (NPFA) as flammable **[44]**. This means that such mixtures with as little as 21 vol % methanol (25 wt %) are flammable liquids; waste methanol in concentrations greater than or equal to 24 % falls under the definition of an ignitable hazardous waste **[62]**. An aqueous solution of 10 % methanol by volume is stated to be stable at atmospheric pressure and a temperature greater than 21 °C (70 °F) **[63]**.

Anecdotal information from Internet chat forums agrees with this basis. Methanol-water mixtures are used as a performance-boosting fuel additive in some cars, and the discussions found center on questions about the maximum concentration for the safe storage, handling, and use of methanol. Methanol fires are particularly dangerous because the flame is invisible in daylight to the naked eye **[21]**. One source relates "kitchen chemistry" experience that a solution with a ratio of 1/3 methanol to 2/3 water cannot be made to catch fire and that a 50 %/ 50 % methanol-water solution ignites but burns poorly **[64]**. Elsewhere, a 50 /50 methanol/water mix by volume is said to be on the edge of flammability **[65]**. An additional source states that a 30 % methanol in water solution just sits there and cannot be lighted in free air **[17]**, and another quotes an industrial brochure as stating that a blend of 33 % methanol / 67 % water is the absolute maximum methanol concentration that will keep the

‡ The autoignition temperature, for which no external ignition source is necessary, is much higher [for example, 686 °F (363 °C) for 100 % ethanol (Ref. **[61]**, p.325-57) and 867 °F (464 °C) for 100 % methanol (Ref. **[61]**, p.325-81)].

mixture non-flammable **[64]**. Still another claims personally to use a 15-20 % methanol solution, which is said to work well, presumably without flammability problems **[64]**.

Finally, flammability tests on methanol-water mixtures **[17]** were conducted outdoors in a metal trough, with results linked to videos posted on other Internet sites. Four methanol concentrations were investigated – 100 %, 50 %, 40 %, and 30 % (all most likely by volume). Combustible paper was used to show up the invisible flame. The burning paper revealed a rather robust combustion for the 100 % case, and the 50 % and 40 % solutions burning with increasing difficulty. The paper from the 30 % trial did not light off at all.

The foregoing observations and findings are consistent with the flash-point entries in Table B4. Flash points of methanol-water solutions are plotted vs. concentration in Ref. **[44]**. Based on the flash-point data, the anecdotal evidence, and the results of the flammability tests, all the calculated methanol-water distillate compositions in Table 2 would likely be flammable when the ambient temperature is about 30 °C (86 °F) or above. In addition, the cumulative samples and most of the individual samples would be expected to ignite at a lower ambient temperature in the laboratory (leaving no residue from combustion of the solution).

Table B4. Flash Points of Methanol-Water Solutions [**44**, **61, 66**]

Methanol Weight %	Methanol Volume %	Temperature	
		°F	°C
0	0	—	—
10	13	130	54
20	24	110	43
30	35	95	35
40	46	85	29
50	56	75	24
60	66	70	21
70	75	60	16
80	83	54	12
90	92	54	12
100	100	54	12

401328 Multicomponent Batch Distillation Optimal Control and Thermodynamic Uncertainties

Abstract:

Batch distillation is commonly used unit operation in specialty chemicals. In order to maximize the profitability and improve separation using batch distillation column, optimal reflux operation can be used. The mathematical and numerical complexities of the optimal control problem get worse when uncertainty is present in the formulation. However, in specialty chemical industries thermodynamic properties of not all chemical components of a mixture to be separated are known. These uncertainties in the problem formulation can result in sub-optimal solutions. These static thermodynamic uncertainties result in dynamic uncertainties due to unsteady state nature of batch distillation operation. In this work, we derive new optimality conditions based on Ito's lemma derived from financial literature where time dependent uncertainties are common. We use stochastic maximum principle for formulation of the resulting stochastic optimal control problem. This algorithm is implemented in the MultiBatchDS batch distillation process simulator. Finally, a numerical case-study derived from a real world batch distillation separation problem encountered at Firmenich is presented to show the scope and application of the proposed approach.

402163 Design, Fabrication and Operation of a Distillation System to Recycle Nuclear Waste

Abstract:

When dealing with nuclear waste, there are a large number of important factors that must be addressed. Many of these are similar to those that should be part of most commercial distillation systems, but differ in degree to which they must be addressed. These start at the early design stage, gathering of process information via pilot plant testing, system design, fabrication, and system operation. Quality control and tracking of all stages of design and construction are a standard requirement.

Some interesting factors that are not normally part of a typical system designs include operating a continuous distillation system on an 8 hours per day schedule, criticality concerns, size constraints due to location of the distillation system in an existing room with thick, shielded concrete walls, materials of construction due to the corrosive mixtures of materials to be processed, and reliability concerns since repair and part replacement become VERY challenging in an high radiation environment.

402165 Novel Methods to Separate Azeotropic Butanol-Water Mixtures without the Need for Extractive Distillation or Pervaporative Membrane Separation

Abstract:

Conventional methods to separate azeotropic mixtures containing isomers of butanol i.e., 2-butanol and tert-butanol, require either multi-column extractive distillation or an integrated pervaporation membrane-distillation configuration to effectively break the azeotropes. While the azeotropic mixture of 2-butanol – water is easy to separate as the aqueous and organic phase compositions lie on the opposite side of the azeotrope, the presence of tert-butanol in the mixture makes the separation complicated as its aqueous and organic phase compositions lie on the same side of the azeotrope, making it difficult to cross the distillation boundary to achieve the required separation.

In this work, various process configurations to effectively bring the mixed butanol product containing 2-butanol and tert-butanol on-specification in-terms of water content were developed without the need for above mentioned cost-intense conventional separation schemes. An optimized separation scheme along with the hydration process was then scaled- up to a 60 kg/day integrated pilot plant and successfully demonstrated the production of on-spec product.

Optimum Control of a Divided Wall Distillation Column

Melissa Donahue
James R. Fair Process Science and Technology Center
The University of Texas at Austin

A basic control structure for a pilot scale divided wall distillation column was commissioned and experimental data was obtained. As part of this effort, temperature pairings for the reflux, top of the dividing wall liquid split, side stream flowrate, and reboiler duty were evaluated to determine the optimal temperature measurement locations for composition control. Wireless temperature transmitters and flow controllers were implemented on a Delta V process control system. As anticipated, the column operation exhibited a high level of control loop interaction. This interaction points to the importance of a variety of process inputs with the liquid split at the top of the dividing wall being the primary control variable. These results will be important in the future as a variety of control strategies and MPC are investigated.

A Total Correlation Evaluating User Fitting UNIFAC Determined from VLE Data Covering 5000 Binaries
(ID#400309)

Satoru Kato[1,2], *Hiroshi Tachibana*[2], *Hidetoshi Kuramochi*[3], *Ichiro Minami*[4]
[1] *TC Lines Co., Ltd., Yokohama, Japan,* [2] *Tokyo Metropolitan University, Hachiohji, Tokyo, Japan,*
[3]*National Institute for Environment Studies, Tsukuba, Japan,* [4] *Epsilon Co., Ltd., Tokyo, Japan*

1. INTRODUCTION

Vapor-liquid equilibria (VLE) are of practical importance for the rational design and control of distillation towers. In the present investigation, a high-precision correlation, called the TC line (Thermodynamic Consistency line)[1-3], is applied to the propylene + propane distillation separations, benzene + cyclohexane + acetone azeotropic distillation and user fitting UNIFAC. The TC line enables us to remove systematic errors from the VLE experimental data. It is a long-standing problem for process simulator users to eliminate inaccurate VLE data predicted using parameters integrated into the process simulators. The TC line is advantageous for eliminating these inaccurate data due to its highly precise convergence. In the following analysis, a total correlation for VLE data covering 5000 binaries[4] is used to evaluate the data consistency of the binary VLE data.

2. MODELS

2.1 Correlation of VLE data using activity coefficient models

The combination of the equation of state and a mixing rule may not satisfy the Gibbs-Duhem equation; therefore, the present investigation uses the activity coefficient models. The vapor-liquid equilibrium equation for component i is formulated as follows [5]:

$$Py_i = \gamma_i x_i p_{si} \quad (i = 1, 2) \tag{1}$$

where P, x_i and p_{si} denote the system pressure, the liquid-phase mole fraction of component i and the vapor pressure of component i, respectively. The activity coefficient, γ_i, is formulated using the following models:

(1) the Margules equation

$$\ln \gamma_1 = x_2^2[A + 2(B - A)x_1], \quad A = \ln \gamma_1^\infty \tag{2}$$

$$\ln \gamma_2 = x_1^2[B + 2(A - B)x_2], \quad B = \ln \gamma_2^\infty \tag{3}$$

(2) the van Laar equation

$$\ln \gamma_1 = A[1 + Ax_1/(Bx_2)]^{-2} \quad A = \ln \gamma_1^\infty \tag{4}$$

$$\ln \gamma_2 = B[1 + Bx_2/(Ax_1)]^{-2} \quad B = \ln \gamma_2^\infty \tag{5}$$

(3) the UNIQUAC equation

$$\ln \gamma_i = \ln \frac{\Phi_i}{x_i} + \frac{z}{2} q_i \ln \frac{\theta_i}{\Phi_i} + \Phi_i \left(l_i - \frac{r_i}{r_j} l_j \right) - q_i \ln(\theta_i + \theta_j \tau_{ji}) + \theta_j q_i \left(\frac{\tau_{ij}}{\theta_i + \theta_j \tau_{ji}} - \frac{\tau_{ij}}{\theta_j + \theta_i \tau_{ij}} \right) \tag{6}$$

$$\Phi_i = \frac{x_i r_i}{x_1 r_1 + x_2 r_2} \tag{7}$$

$$\theta_i = \frac{x_i q_i}{x_1 q_1 + x_2 q_2} \tag{8}$$

$$l_i = \frac{z}{2}(r_i - q_i) - (r_i - 1) \quad (9)$$

$$\ln \tau_{12} = -\frac{u_{12}}{RT} \quad (10)$$

where the molecular volume, r_i, and the molecular surface area, q_i, of component i are given in ref. 6. In Eq. (10), R and T denote the gas constant and system temperature, respectively.

2.2 An Example of VLE data correlation

In **Fig. 1**, P is plotted versus x_1 at 373.15 K for the methanol (1) + water (2) binary. Any of the activity coefficient equations can satisfactorily correlate the non-azeotropic mixtures, from which the γ_i^∞ values can be determined.

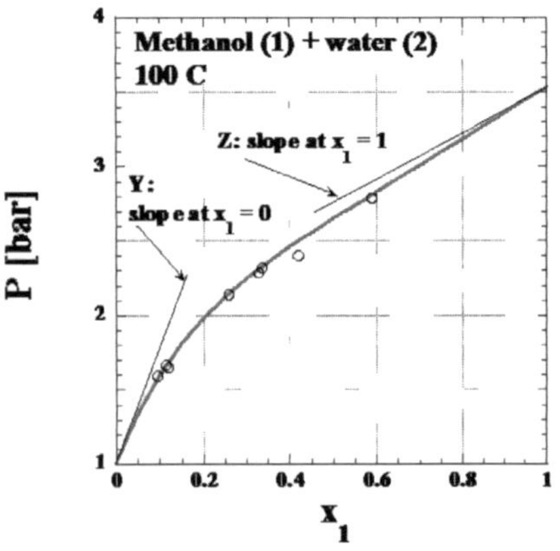

Fig. 1 P vs. x_1 at 373.15 K for the methano (1) + water (2) binary system, (o) experimental data[7], (—) correlated using the Margules equation with $A = 0.882$ and $B = 0.539$.

2.3 Faults in conventional correlations

In **Fig. 2**, the UNIQUAC binary parameter, u_{12}, is plotted versus $T - 273.15$ for the methanol (1) + water (2) binary. In **Fig 3**, the margules binary parameter, A, is plotted versus $T - 273.15$. These figures show that data scatterings are notable, and a line representing the data cannot be fixed. Similar faults occour even if other activity coefficient equations are used. Furthermore, the scattering is more notable, if azeotropic mixtures are analyzed.

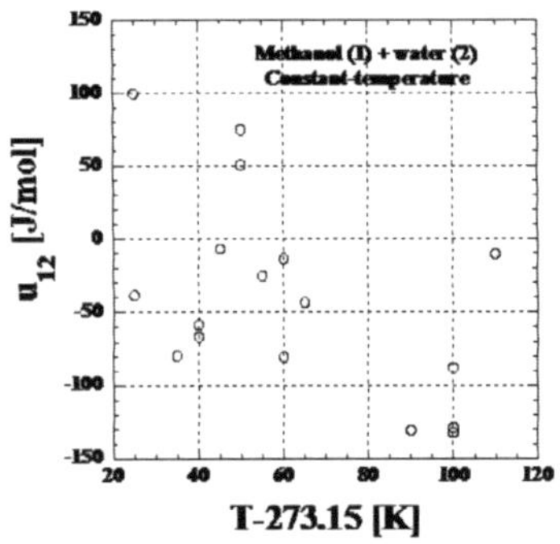

Fig. 2 u_{12} vs. $T - 273.15$ for the methanol (1) + water (2) binary, (o) UNIQUAC parameters calculated from constant temperature data[7-15].

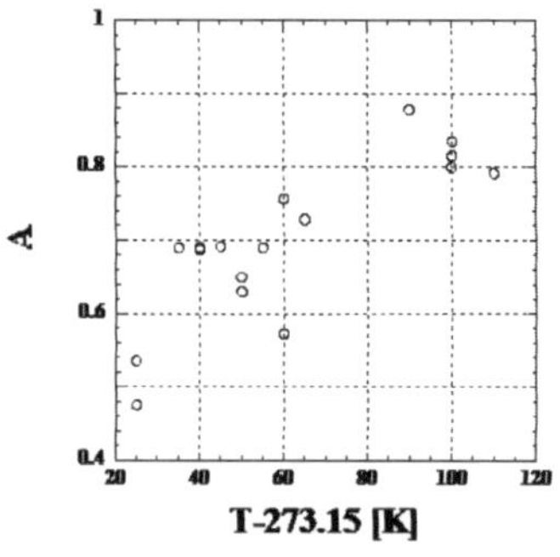

Fig. 3 A vs. $T - 273.15$ for the methanol (1) + water (2) binary, (o) Margules parameters calculated from constant temperature data[7-15].

3. TC LINE

3.1 Definition and advantages of TC line

The TC line is high-precision correlations that occur between X versus Y and X versus Z. The quantities, X, Y and Z, are defined as follows[2]:

$$X = (p_{s1} + p_{s2})/(P_{c1} + p_{s2}) \quad (11)$$

$$Y = (p_{s1}\gamma_1^{\infty} - p_{s2})/(P_{c1} - p_{s2}) \quad (12)$$

$$Z = (p_{s1} - p_{s2}\gamma_2^{\infty})/(P_{c1} - p_{s2}) \quad (13)$$

where X and P_{c1} denote the proximity ratio to the critical point and critical pressure, respectively. As shown in Fig. 1, Y and Z denote the non-dimensional infinite dilution pressure gradients at x_1=0 and 1, respectively.

Examples of the TC lines are shown in **Figs. 4 a and b**. In Fig. 4, the following X, Y and Z were used, because, at a low pressure VLE, the effect of the denominators in Eqs. (11)-(13) are negligible.

$$X = (p_{s1} + p_{s2}) \quad (14)$$

$$Y = (p_{s1}\gamma_1^{\infty} - p_{s2}) \quad (15)$$

$$Z = (p_{s1} - p_{s2}\gamma_2^{\infty}) \quad (16)$$

The TC line is formulated as Eq. (17) or Eq. (18).

$$Y, Z = aX^b \quad (17)$$

$$Y, Z = aX + bX^2 + c\ln(X+1) \quad (18)$$

The great advantages of the TC line are that it passes through the origin, and constant-temperature VLE data and constant-pressure VLE data converge to the same line, as shown in Figs. 4 (a) and (b).

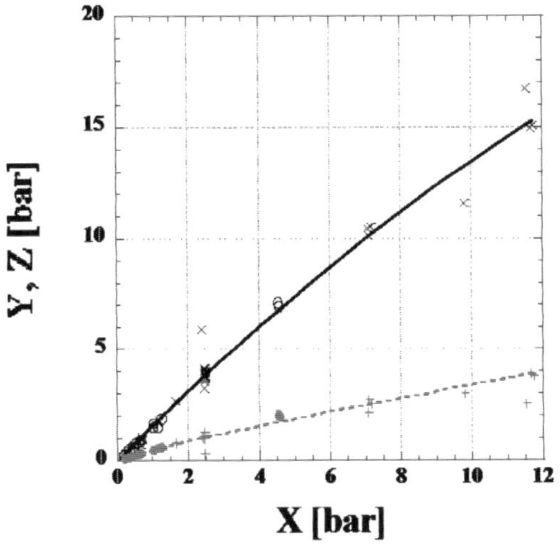

(a)

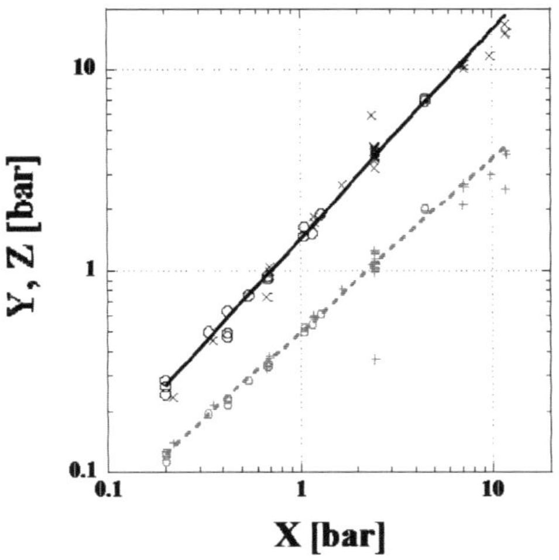

(b)

Figs. 4 (a) and (b) Y and Z vs. X for the methanol water bianry, (o) X vs. Y calculated from constant temperature data, (x) X vs. Y calculated from constant-pressure data, (—) X vs. Y TC line with $Y = 1.73X^{1.04}$ as shown in ref. 3, (●) X vs. Z calculated from constant-temperature data, (+) X vs. Z calculated from constant pressure data, (- - -) X vs. Z TC line with $Z = 0.283X^{0.871}$ as shwon in ref. 3; exp. data were cited from ref. 6

3.2 *x*, *y* Relationships predicted from the TC line

In **Fig. 5**, y_1 is plotted versus x_1 for the benzene (1) + 1-propanol (2) binary at 1.013 bar, in which the van Laar equation was used for the prediction of y_1. The van Laar parameters were determined using the TC lines shown in **Fig. 6.** In **Fig. 7**, y_1 is plotted versus x_1 for the 2-butanone (1) + water (2) binary at 1.013 bar. The van Laar equation was also used, and the van Laar parameters were determined using the TC lines shown in **Fig. 8**. Figs. 5 and 7 show that accurate *x*, *y* relationships are predicted using the TC line, even if the azeotropic and the two-liquid phase forming binaries are analyzed.

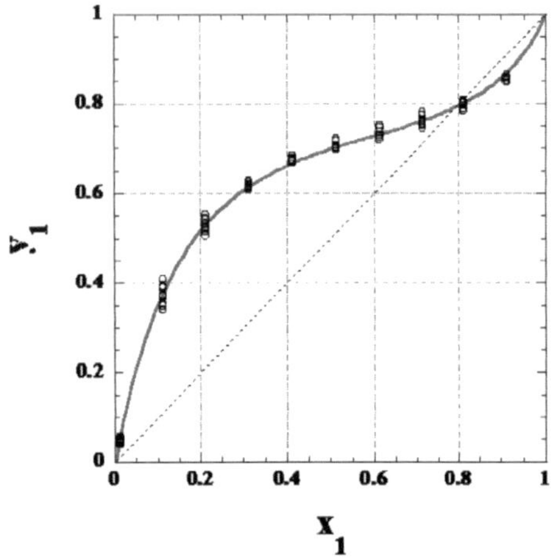

Fig. 5 y_1 vs. x_1 for the benzene (1) + 1-proapnol (2) binary at 1.013 bar, (o) exp. data, (—) predicted using TC lines.

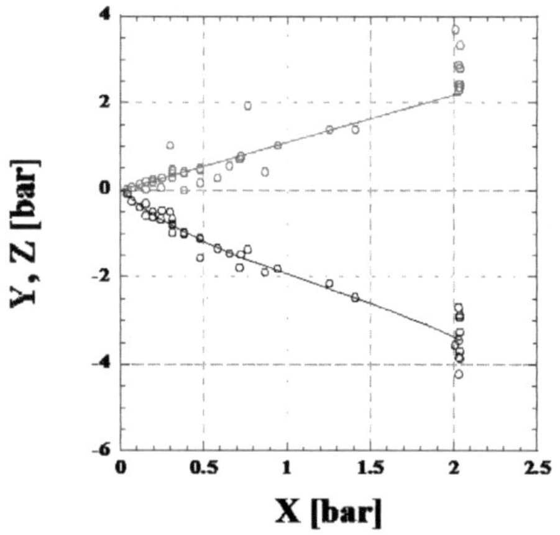

Fig. 6 *Y* and *Z* vs. *X* for the benzene + 1-propanol binary, (o) *X* vs. *Y* calculated from exp. data, (—) *X* vs. *Y* TC line, (o) *X* vx. *Z* calculated from exp. data, (—) *X* vs. *Z* TC line, exp. data were cited from ref. 6.

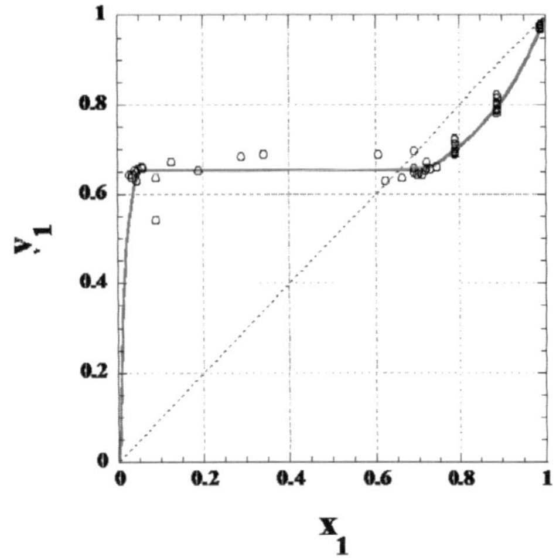

Fig. 7 y_1 vs. x_1 for the 2-butanone (1) + water (2) binary at 1.013 bar, (o) exp. data, (—) predicted using TC lines.

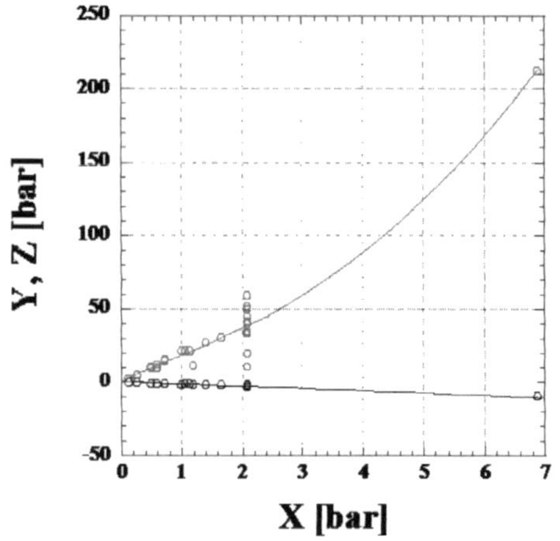

Fig. 8 *Y* and *Z* vs. *X* for the 2-butanone + water binary, (o) *X* vs. *Y* calculated from exp. data, (—) *X* vs. *Y* TC line, (o) *X* vx. *Z* calculated from exp. data, (—) *X* vs. *Z* TC line, exp. data were cited from ref. 6.

4. TOTAL CORRELATION

In **Fig. 9**, I_{exp} is plotted versus I_{cal} using the TC lines satisfying $n \geq 7$, $AADY < 0.2$ and $AADZ < 0.2$, where n denotes the number of data sets, and *AADY* and *AADZ* are the average deviations from the *Y* and *Z* TC lines, respectively. The total correlation function, *I*, includes universal constants, which are independent of the system. Fig. 9 shows that $I_{exp} = I_{cal}$ holds, if reliable TC line data are used. In **Fig. 10**, I_{exp} is plotted versus I_{cal} for 5000 binaries. Of the 5000 binaries, 92 % are reliable, because they satisfy the relationship, $I_{exp} = I_{cal}$. Most of the binaries reflecting low quality data have only one dataset.

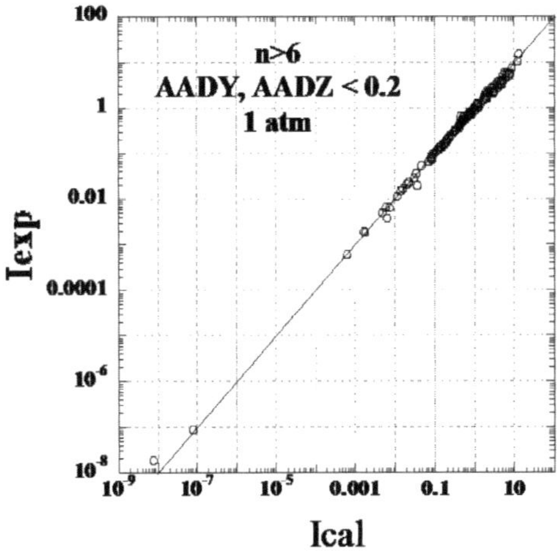

Fig. 9 I_{exp} vs. I_{cal}, plotted using reliable TC line data, as $n \geq 7$, $AADY < 0.2$ and $AADZ < 0.2$.

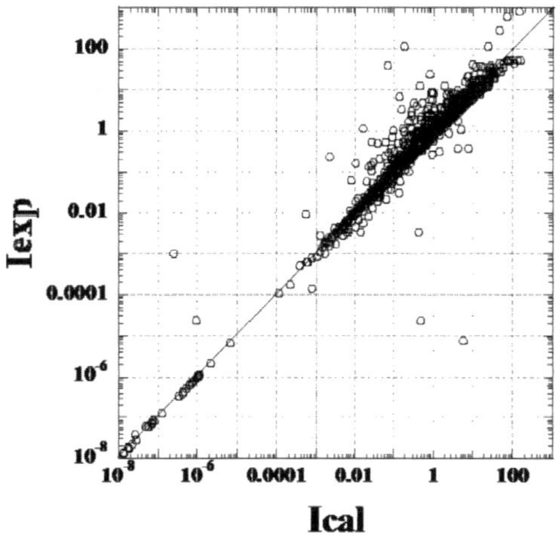

Fig. 10 I_{exp} vs. I_{cal}, plotted using the TC line data of 5000 binaries.

5. APPLICATIONS

5.1 Propylene + propane distillation separation

To show the usefulness of the TC line and the total correlation, the number of theoretical stages operated at total reflux satisfying three-nine purities was calculated using the Ponchon-Savarit method. The method of Rowlinson was used for the prediction of the pure liquid heat capacities. The partial molar excess enthalpies were calculated from the temperature dependence of the TC line. This procedure is a notable advantage that originated in the TC line. In **Fig. 11**, the TC lines for the propylene + propane binary are plotted. The infinite dilution activity coefficient values were calculated from the Margules equation determined from the *P*, *x* data, as shown in **Fig. 12**. As shown in **Table 1**, the number of theoretical stages, *N*, calculated at $T = 223.75$ K was 80. The TC lines at 223.75 K satisfy the total correlation; therefore, they are reliable. Meanwhile, the experimental data at 223.75 K[16], shown in Fig. 11 using the symbols ▼ and ▲, resulted in $N = 66$. Conventional consistency tests cannot eliminate inaccurate data. In contrast to this fault, the TC line can provide accurate VLE.

Table 1 Number of theoretical stages at total reflux satisfying three-nine purities.

System	Purpose	*N* from TC line	*N* from exp. data
Propylene + propane	C3 distillation	80	66
Benzene + cyclohexane + acetone	Azeotropic distillation	17	15
Benzene + cyclohexane + acetone	User fitting UNIFAC	17	15

N: number of theoretical plates calculated using the Ponchon-Savarit method.

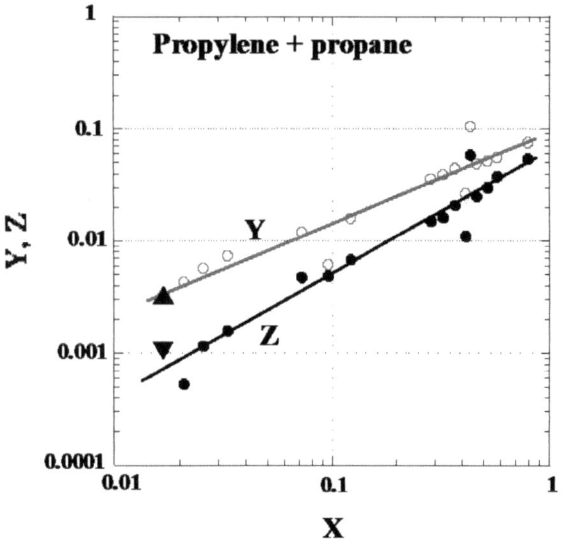

Fig. 11 TC lines for the propylene (1) + propane (2) binary, where $T < T_{c1}$ is satisfied.

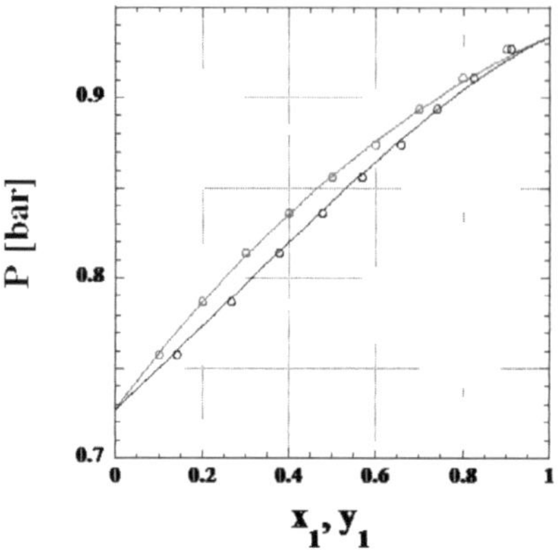

Fig .12 P vs. x_1 and y_1 for the propylene (1) + proapne (2) binary at 223.75 K, (o) P vs. x_1 data[16], (o) P vs. y_1 data[16], (—) correlation of P, x data using the Margules equation, and (—) P, y predicted using the Margules equation.

5.2 Azeotropic distillation and an application to model predictive control of distillation towers

Benzene is separated from a mixture with cyclohexane by azeotropic distillation with acetone. The TC lines of the three binaries consisting of the ternary system, benzene + cyclohexane + acetone, satisfy the total correlation. The UNIQUAC equation was used for the calculation of the activity coefficients in the ternary system. As shown in Table 1, $N = 17$ was obtained using the TC lines. Meanwhile, if the TC lines of one of the binaries are replaced with experimental data, the worst case was $N = 15$. As shown in **Fig. 13**, the partial molar excess enthalpy cannot be determined from a conventional correlation. Meanwhile, the TC lines, as shown in **Fig. 14**, can provide an accurate liquid phase enthalpy. In **Fig. 15**, (L/Q) divided by its value at the bottom is plotted versus N, where L and Q denote the liquid flow rate and heat load, respectively. Fig. 15 shows that, by using the TC line method, McCabe-Thiele and athermal approximations, respectively, provide 3% and 2% differences in the (L/Q) values from that calculated using the TC lines. It is estimated that the difference increases with the increasing gap in solute volatility. The TC line method will be integrated into a process simulator using a pseudo binary method (PBM)[17, 18] for calculating the real-time model predictive control of distillation towers.

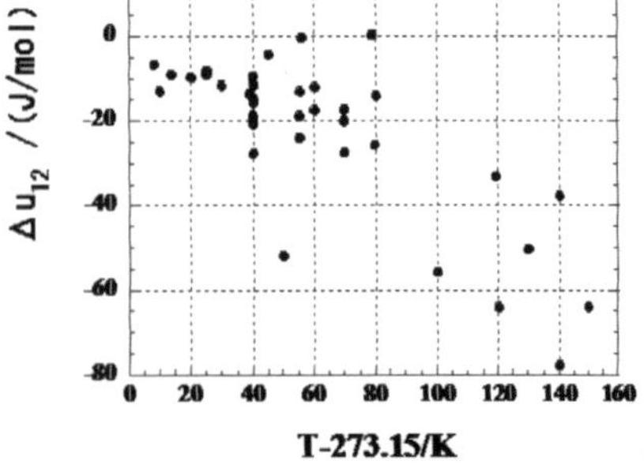

Fig. 13 u_{12} vs. $T - 273.15$ for the benzene (1) + cyclohexane (2) binary, (●) exp. data determined using the UNIQUAC equation.

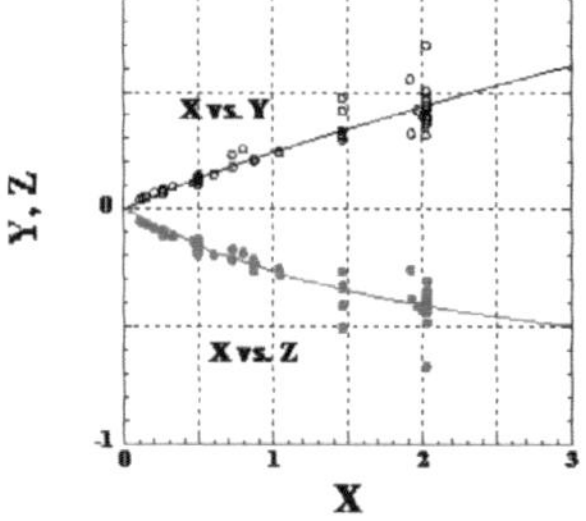

Fig 14. Y and Z vs. X for the benzene (1) + cyclohexane (2) binary, (o) X vs. Y calculated from exp. data using the UNIQUAC equation, (—) X vs. Y TC line with $Y = 0.239X^{0.831}$, (●) X vs. Z calculated from exp. data using the UNIQUAC equation, (—) X vs. Z TC line with $Z = -0.255\ X^{0.692}$.

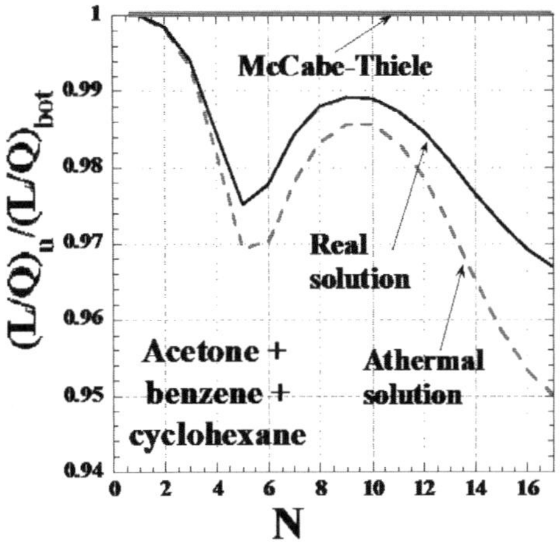

Fig. 15 $(L/Q)/(L/Q)_{bot}$ vs. N for the benzene + cyclohexane + acetone system, (—) McCabe-Thiele, (- - -) athermal solution, and (▬) real solution calculated from TC lines.

5.3 User fitting UNIFAC

a. Establishment of group interaction parameters

UNIFAC group interaction parameters can be obtained from the Dortmund Data Bank. However, accurate UNIFAC parameters can be determined using the TC lines by user's own fitting, because experimental errors have been eliminated from the TC lines. To perform this, the UNIFAC group interaction parameters can be optimized using the following objective function:

$$O.F. = (1/n)\Sigma|(\ln\gamma_i^{\infty}{}_{TC} - \ln\gamma_i^{\infty}{}_{UNI})/\ln\gamma_i^{\infty}{}_{TC}| \quad (19)$$

Interaction parameters between the CH_3OH and $ACNO_2$ groups are lacking in the original UNIFAC (ver. 5)[5]. In **Table 2**, optimized parameters are listed determined from the user fitting UNIFAC using the TC lines of the methanol + nitrobenzene binary. The two parameters can be determined using the two infinite dilution activity coefficient values of methanol and nitrobenzene.

Table 2 Interaction parameters, a_{nm} and a_{mn}, for the methanol + nitrobenzene binary

Group pair (m, n)	a_{mn}	a_{nm}
CH_3OH - $ACNO_2$	139.1 K	978.7 K

b. Prediction of VLE for binaries lacking experimental data

In **Fig. 16**, the x, y relationship for the water (1) + 4-methylpyridine (2) binary is predicted. The UNIFAC group interaction parameters were optimized using the TC lines of the following six binaries: pyridine +, 2-methylpyridine +, 3-methylpyridine +, 2,6-dimethylpyridine +, 2,4-dimethylpyridine+ and 2-methyl-5-ethylpyridine + water. In **Fig. 17**, y_1 is plotted versus x_1 for the acetone (1) + dodecane (2) binary. Five acetone + alkane binaries were used except for the acetone + dodecane binary. Figs. 16 and 17 show that the user fitting UNIFAC is useful for the prediction of VLE for a binary, in which experimental data are lacking.

c. Improvement of group interaction parameters

Accurate group interaction parameters can be determined from the user fitting UNIFAC using appropriate TC lines. In **Fig. 18**, x, y relationships for the acrolein + water binary is plotted, as determined using the TC lines of five binaries: acrolein + water, 1-butanol + octen-4-al, 1-pentene + acetaldehyde and crotonaldehyde + vinyl acetate. In **Fig. 19**, y_1 is plotted versus x_1 for the vinyl acetate (1) + chrotonaldehyde (2) binary. Figures 18 and 19 show that the user fitting UNIFAC is useful for improving group interaction parameters to be accurate. The success of the user fitting UNIFAC suggests the possibility of another user fitting group contribution model using the TC lines and the NRTL or van Laar equation accompanied with SMILES code group assignment.

d. Strength of molecular interactions acting on chemical groups

The strength of molecular interactions acting on a chemical group is formulated by the partial molar excess enthalpy of the group, H_i, as follows:

$$H_i = \frac{\partial \ln \Gamma_i}{\partial (1/T)} \qquad (20)$$

In Eq. (20), Γ_i denotes the activity coefficient of chemical group i. In **Table 3**, partial molar excess enthalpy, H_i/RT, and entropy, S_i/R, of CH_2=CH group in the acrolein + water binary and CH_3 group in the 1-butanol +water binary are shown as calculated using the TC lines at 333.15 K. Table 3 clearly shows that the CH_2=CH group is instable in water, because $\Delta H_i/RT >> 0$ and $\Delta S_i/R >> 0$ hold. One of the great advantages of the TC line is that it can provide the accurate temperature dependence of Γ_i., from which the strength of molecular interactions can be quantitatively identified. This fact can be applied to chemical and pharmaceutical material development procedures.

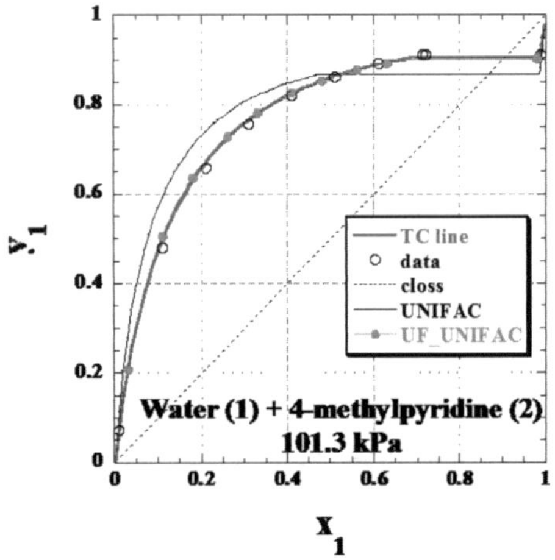

Fig. 16 y_1 vs. x_1 for the water (1) + 4-methylpyridine (2) binary at 1.013 bar, (o) exp. data, (—) predicted using TC lines, (—) predicted using UNIFAC (Ver. 5), and (—•—) predicted using the user fitting UNIFAC.

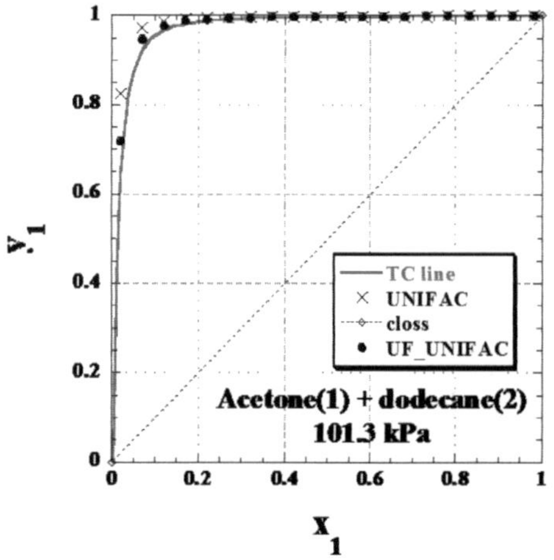

Fig. 17 y_1 vs. x_1 for the acetone (1) + dodecane (2) binary at 1.013 bar, (—) predicted uaing TC lines, (x) predicted using UNIFAC (Ver. 5), and (•) predicted using the user fitting UNIFAC.

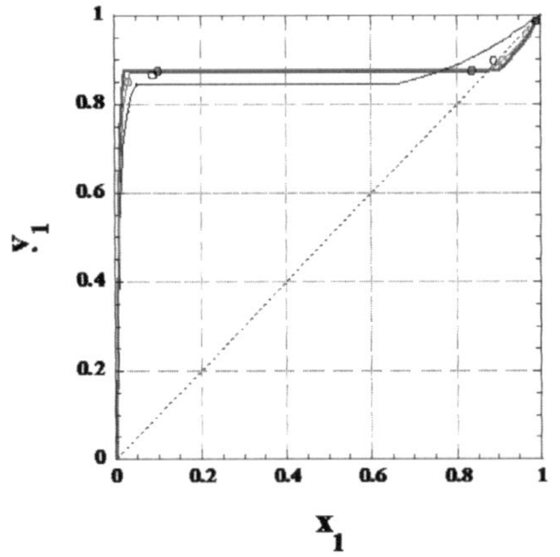

Fig. 18 y_1 vs. x_1 for the acrolein (1) + water (2) binary at 1.013 bar, (o) exp. data, (—) predicted using TC lines, (—) predicted using UNIFAC (Ver. 5), and (—•—) predicted using the user fitting UNIFAC with optimized parameters.

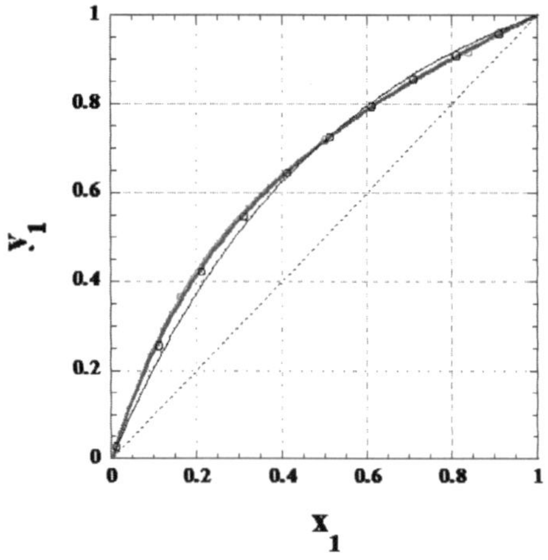

Fig. 19 y_1 vs. x_1 for the vinyl acetate (1) + croton aldehyde (2) binary at 1.013 bar, (o) exp. data, (—) predicted using TC lines, (—) predicted using UNIFAC (Ver. 5), and (—•—) predicted using the user fitting UNIFAC with optimized parameters.

Table 3 Infinite dilution partial molar excess enthalpies and entropies of chemical groups

	$H_i/333.15R$	S_i/R		$H_i/333.15R$	S_i/R
CH_2=CH in water	6.6	3.9	CH_3 in water	-0.83	-2.4
CH_2=CH in acrolein	-0.27	-0.6	CH_3 in 1-butanol	-0.13	-0.4
H_2O in acrolein	2.3	0.33	H_2O in 1-butanol	0.2	-1.1

6. CONCLUSION

The TC line accompanied with the total correlation is useful for predicting the accurate number of theoretical stages of distillation towers. It was shown that the user fitting UNIFAC can predict accurate VLE, even if the VLE data are lacking. Also, the user fitting UNIFAC can improve group interaction parameters. Its applications to material development procedures and SMILES code grouping are expected.

Acknowledgement

The authors are grateful to the Tokyo Metropolitan University for the approved use of part of the TC lines applied for patents.

References

1 Patent: US20140288905A1, EP276218A1, KR20140079425A, CN103906553A, WO2013047494A1.
2 S. Kato, J. P. Schmelzer, P. Gostomski, K. N. Marsh, J. Chem. Eng. Data, 56 (2011) 4927-4934.
3 S. Kato, P. Gostomski, A. Oda, J. Freitag, S. Hirohama, D. Bluck, Fluid Phase Equilib., 357 (2013) 36-42.
4 S. Kato, Kagaku Kogaku, 77 (2013) 474-477, http://tc-lines.com
5 B.E. Poling, J.M. Prausnitz, J.P. O'Connell, The Properties of Gases and Liquids, 5th ed. McGraw-Hill, New York, 2001.
6 J. Gmehling, U. Onken, Vapor–Liquid Equilibrium Data Collection, Chemistry, Data Series, vol. 1, parts 1–8, 1981, Dechema.
7 M. Albert, I. Hahnenstein, H. Hasse, G. Maurer, AIChE J., 42 (1996) 1741-1752.
8 M.L. MC Glashan, A.G. Williamson, J. Chem. Eng. Data, 21 (1976) 196-199.
9 G.A. Ratcliff, K.C. Chao, Can. J. Chem. Eng., 47 (1969) 148-153.
10 M.S.H. Bader, K.A.M. Gasem, Chem. Eng. Commun., 140 (1995) 41-72.
11 K. Kurihara, T. Minoura, K. Takeda, K. Kojima, J. Chem. Eng. Data, 40 (1995) 679-684.
12 M.J. Lee, S.L. Chen, C.H. Kang, H.M. Lin, Ind. Eng. Chem. Res., 39 (2000) 4383-4391.
13 P. Sentenac, Y. Bur, E. Ruzy, C. Berro, J. Chem. Eng. Data, 43 (1998) 592-600.
14 J. Yao, H. Li, S. Han, Fluid Phase Equilib., 162 (1999) 253-260.
15 J. Zielkiewicz, P. Oracz, Fluid Phase Equilib., 59 (1990) 279-290.
16 M. Hirata, T. Hakuta, T. Onoda, Int. Chem. Eng., 8 (1968) 175.
17 I. Minami, Kagaku Souchi, (2003) 92.
18 Y. Ishii, F.D. Otto, Compt. Chem. Eng. ,32 (2008) 1842.

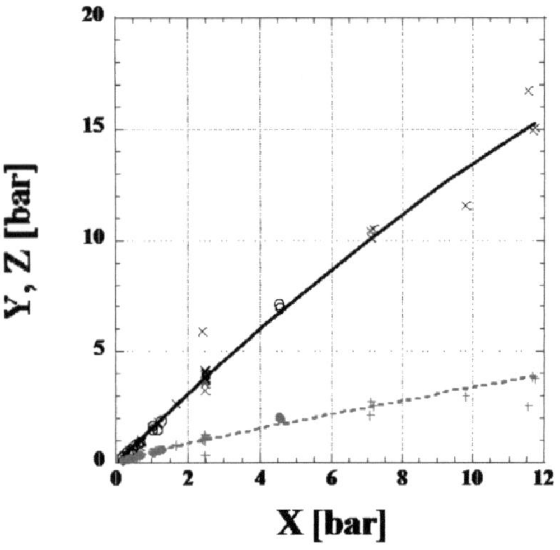

(a)

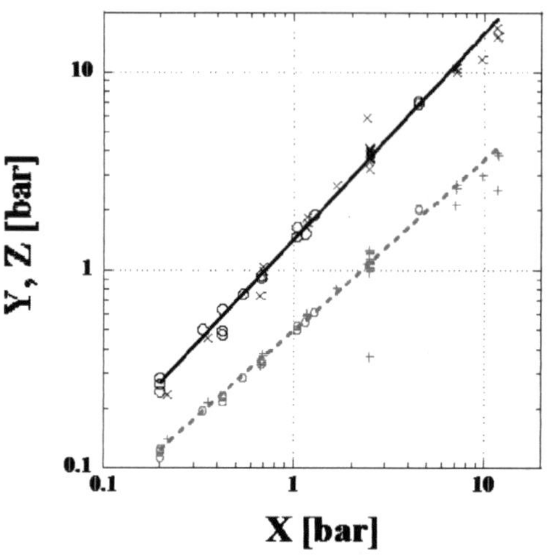

(b)

MODIFICATION OF THE BILLET AND SCHULTES-MODEL FOR THE HYDRAULIC MODELLING OF RANDOM AND STRUCTURED PACKING

Paper 4a

Verena Wolf, Markus Lehner, Montanuniversitaet Leoben, Leoben, Austria
Karin Hoffmann, RVT Process Equipment GmbH, Steinwiesen, Germany

Abstract

If a new packing is developed and introduced to the market, a reliable calculation of the properties is hard to perform as the known correlations are not able to consider all the special influences of the geometrical properties of a packing. Therefore, the published hydraulic models only work acceptable for validated packing types.

To ensure a good rating of a newly developed packing for different application areas, a reliable forecast of the hydraulic properties has to be guaranteed. Therefore, the following models from literature have been examined on their ability to model a newly developed grid-structured packing: SRP-Model, Delft-Model, NNA-Model of Piché et al., Billet and Schultes-Model, Maćkowiak-Model and Stichlmair-Engel-Model. Unfortunately, none of them is able to calculate the hydraulic properties of the grid structured packing satisfactorily. The best modelling results could be reached with the model of Billet and Schultes. As a result of that and the fact that the Billet and Schultes-Model is applicable for structured as well as for random packing, it was examined more in more detail. Due to a relatively simple but physically justified modification of the Billet and Schultes-Model good modelling results can be reached.

Furthermore, it can be shown that this adjustment also leads to a better prediction of the wet pressure drop of further examined packing types. Moreover, the area of validity of the Billet and Schultes-Model of 0.61 – 60 $m^3/(m^2*h)$ can be extended to an irrigation density of 120 $m^3/(m^2*h)$ for the calculation of the wet pressure drop.

Introduction

The historical development of random and structured packings is characterised by a continuous improvement of the hydraulic capacity as well as the mass transfer properties. Already decades ago it was ascertained that a structured arrangement of the packing is highly advantageous in order to reduce the pressure loss. Due to the closed structure of conventional plate packings, the exchange of gas and liquid in horizontal direction is limited. Furthermore, particularly in terms of plastic structured packings, the production costs are high compared with random plastic packings. RVT Process Equipment developed a new grid structured plastic packing, called Hiflow Plus (Figure 1). This packing is optimized for the material consumption and the open structure enables a free horizontal exchange of gas and liquid. This new packing type aims to close the gap between random and structured packings.

The fields of industrial application of this grid structured packing are separation processes with high volume flow rates and moderate temperatures as for example sea water scrubbers for SO_2 removal from exhaust gases of power plants, amine scrubbers for carbon capture units or ground water strippers, just to mention a few. (1)

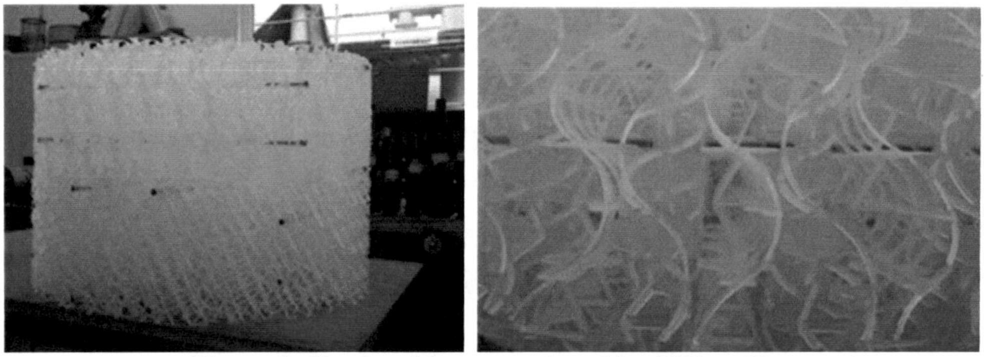

Figure 1: Plastic grid structured packing Hiflow Plus

For the introduction of a newly developed packing to the market, a reliable calculation of the hydraulic properties is needed. With the known correlations from literature this is hard to perform as these models are not able to consider all the special influences of the geometrical properties of a packing. Consequently, the published hydraulic models only work acceptable for validated packing types. Therefore, detailed and reliable experimental measurements are necessary in order to validate the results of existing models, or to develop a new correlation for the calculation of the packing properties.

Experimental Set-Up

The hydraulic properties (dry and wet pressure drop, hold-up, loading and flooding point) have been examined at an experimental plant of the Chair of Process Engineering of Industrial Environmental Protection at Montanuniversitaet Leoben. In Figure 2 the flow sheet of the used experimental set-up is depicted. The plant consists of a DN 600 saturation column and a DN 450 measurement column, both made of polypropylene. The air is sucked from the environment and flows through the saturation column where it is saturated with water. After passing a tube type gas distributor, the gas streams through the measurement column. Afterwards, it enters a radial blower (frequency converted gear, max. 5000 m^3/h, max. 5000 Pa pressure increase) and is then finally released to the environment. The gas flow is measured by a flow grid (airflow) downstream of the blower. Furthermore, the temperature of the gas is measured at every in- and outlet of both columns (Endress & Hauser, PT 100 type Omnigrad M TR 13; WIKA, type 55) which have their own closed liquid circulation. The packing is irrigated by a trough type liquid distributor (128 dripping points per m^2, free cross sectional area of 41%, three available operating ranges: 5-30 $m^3/(m^2h)$, 5-60 $m^3/(m^2h)$ and 30-120 $m^3/(m^2h)$). The used radial pumps are able to handle a volume flow of 50 m^3/h. The liquid flow is measured by a magnetic inductive flow meter (Krohne, type optiflux 4000) and the temperature of the inlet stream at the measurement column is gauged by a resistance thermometer (Voltcraft, type K).

The hydraulic properties of a packing are examined at different gas and liquid loads. At the beginning of a test run the adjustments of the selected loads are made and it is necessary to wait until steady state conditions have been reached. After five minutes of stable operation all necessary values can be read off. Then the liquid level in the bottom tank of the measuring column is raised to a defined height by opening the water supply line which is necessary for the hold-up measurement described below. Afterwards, the pumps and the blower are shut down simultaneously. During the following 30 minutes the liquid can drop off from the packing completely and leave the column bottom tank before the next operating point (same irrigation density but higher F-factor) can be adjusted. Once the flooding point for the chosen irrigation density is reached, the test series can be accomplished for the next irrigation density.

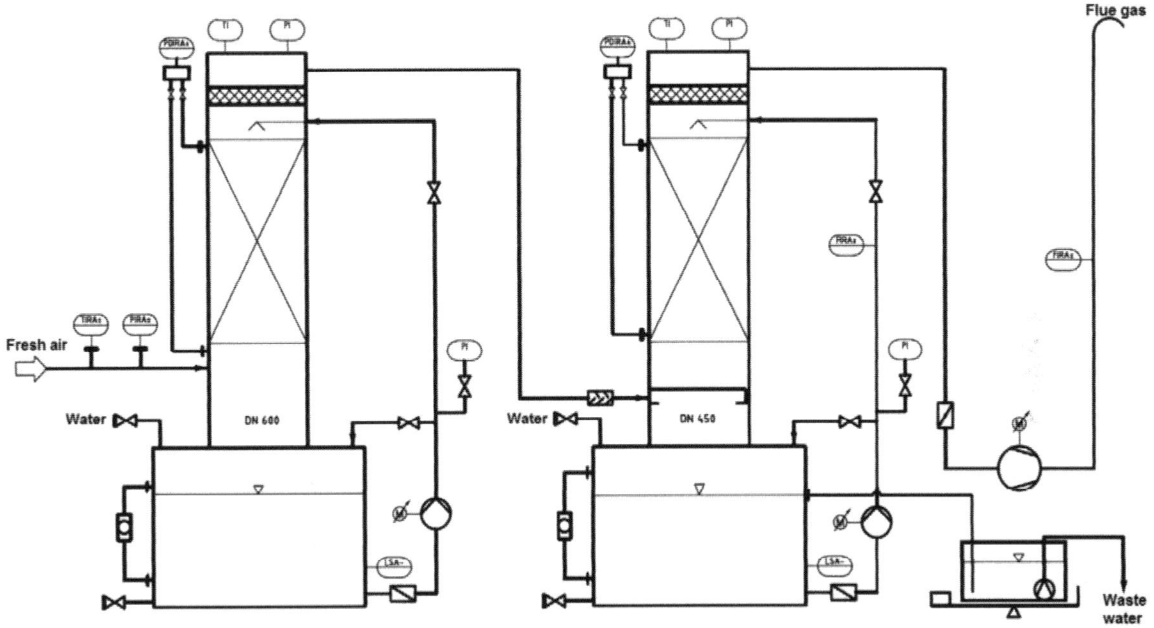

Figure 2: Flow sheet of the experimental plant "Packed column"

Pressure Loss Measurement

The pressure loss can be examined by calculating the difference of the pressure measurements which are taken by a differential pressure transmitter (Endress & Hauser, Deltabar S PMD75) situated at the upper and the lower region of the packing. Due to the experimental set up the static pressure is measured at these positions, and therefore it needs to be ensured that the dynamic pressure is equal on both pressure measurement positions. This can be guaranteed if the gas velocity is constant wherefore an equal available cross-section for the gas stream is required. Therefore, both pressure measurement positions have to be either located "inside" of the packing, or below and above of the packing. To be able to neglect the influence of the support grid and a hold-down grid, which is used in some cases, the pressure measurement positions are located at a column height where packing is situated in the column. For the design of the pressure measuring taps the findings of Cai and Resetarits (2) were considered.

Hold-up Measurement

The measurement of the hold-up is achieved by collecting and weighing the amount of water which is in the column during operation. Therefore, the water level in the bottom tank of the measuring column is adjusted by adding water from the tap to a defined height during operation. This height is given by the lower end of a nozzle in the tank, from which a pipe is leading to a tub standing on a scale

(Figure 3). Afterwards, the column is shut down and the operating liquid is flowing out of the bottom tank into the tub on the scale, where it can be weighed. To exactly calculate the hold-up of the packing itself, a calibration of the column without packing is needed to be able to subtract the amounts of liquid which are situated in the pipe, in the liquid distributor and as droplets below and above of the packing (Figure 3).

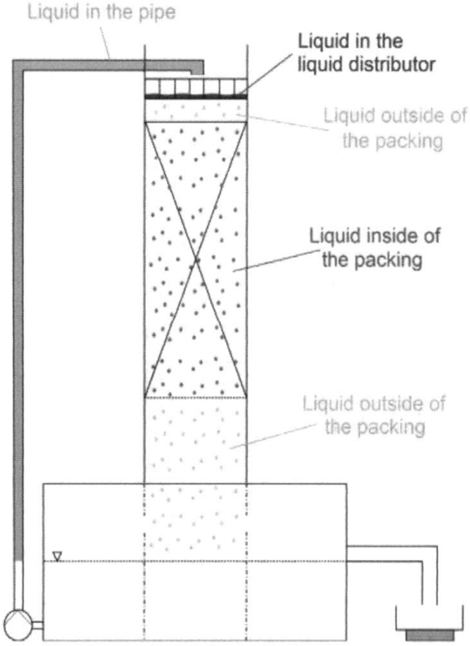

Figure 3: Setup for hold-up measurements with different parts of liquid situated in the column during operation

Loading and Flooding Point Measurement

The loading point can be identified by the change of the slope of the pressure loss line, when the pressure loss is plotted over the F-factor in a double logarithmic diagram. For this reason the measurement results for each operating point are immediately entered into a spread sheet including such a diagram. Furthermore, the increase of the gas load is done in smaller steps after reaching the loading point for a special irrigation density to ensure a better resolution of the hydraulic properties in the loading zone.

The flooding point can be detected if it is not possible to reach stationary conditions. At this point the pressure loss over the packing is not stable any more but is increasing although the gas and the liquid loads are constant. Furthermore, the flooding point can also be identified in the above mentioned diagram when the slope of the pressure loss line approaches the value of one.

Hydraulic Modelling

Following models from literature have been examined on their ability to rate the newly developed grid-structured packing: SRP-Model (3-10), Delft-Model (10-12), NNA-Model of Piché et al. (13-18), Billet and Schultes-Model (19-28), Maćkowiak-Model (29-31) and Stichlmair-Engel-

Model (32-34). Unfortunately, none of them is able to calculate the hydraulic properties of the grid structured packing satisfactorily. The best modelling results could be reached with the model of Billet and Schultes. As a result of that, and the fact that the Billet and Schultes-Model is applicable for structured as well as for random packing, it was examined in more detail.

Billet and Schultes-Model

The Billet and Schultes-Model assumes the packing as a bundle of parallel channels, and that the liquid trickles down as a laminar flowing film with constant film thickness in the inside of these channels while the gas phase flows upwards in counterflow. By neglecting the inertia forces and assuming that the liquid film is in equilibrium state, gravity and shear forces in the liquid film are held in equilibrium with the frictional forces by the shear stress in the vapour phase at the surface of the film (Figure 4). (23)

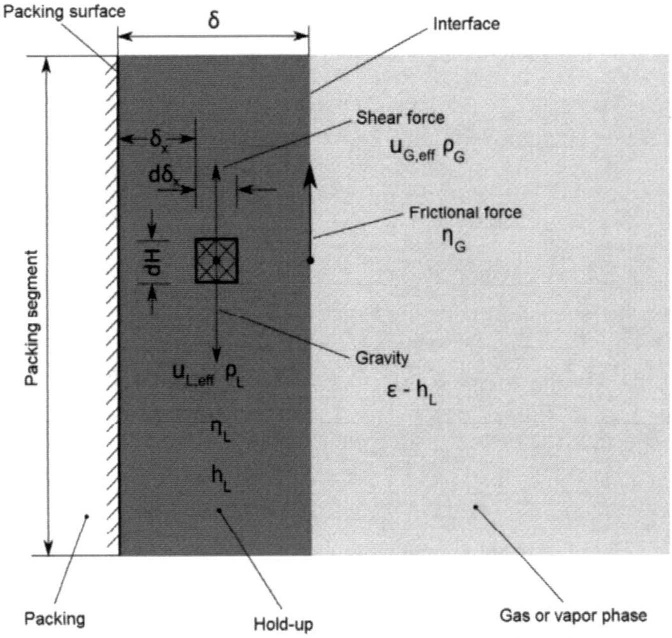

Figure 4: Theoretical assumptions of the Billet and Schultes-Model (26)

Based on this theoretical background the equations for the calculation of the loading and the flooding point as well as the liquid hold-up **at** respectively **below** of the loading point are established. The correlation for the hold-up at the flooding point is derived from empirical observations and the calculation equation for the hold-up in the loading zone is an interpolation of the 13th order between the hold-up at the loading point and the one at the flooding point. The calculation of the dry pressure drop (Equations (1)-(3)) is based on the Ergun-Equation (35) which is in turn a further development of the Darcy-Weisbach Equation.

$$\frac{\Delta p_{tr}}{H} = \psi_{tr} * \frac{a_{geo}}{\varepsilon^3} * \frac{F^2}{2} * \frac{1}{K} \tag{1}$$

$$\frac{1}{K} = 1 + \frac{2}{3} * \frac{1}{1-\varepsilon} * \frac{d_P}{d_S} \tag{2}$$

$$\psi_{tr} = C_{P,0} * \left(\frac{64}{\text{Re}_G} + \frac{1,8}{\text{Re}_G^{0,08}} \right) \qquad (3)$$

The wet pressure drop (Equation (4)) is again calculated similar to the dry pressure drop but the herein used resistance coefficient is combined by four terms (Equation (5)).

$$\frac{\Delta p}{H} = \psi_L * \frac{a_{geo}}{(\varepsilon - h_L)^3} * \frac{F^2}{2} * \frac{1}{K} \qquad (4)$$

$$\psi_L = \psi_{tr} * \left(\frac{\varepsilon - h_L}{\varepsilon} \right)^{1,5} * \left(\frac{h_L}{h_{L,S}} \right)^{0,3} * B_f \qquad (5)$$

- Dry resistance coefficient ψ_{tr},
- one term, who takes into account that the free cross sectional area is reduced through the down trickling fluid $\left(\frac{\varepsilon - h_L}{\varepsilon} \right)^{1,5}$,
- the term $\left(\frac{h_L}{h_{L,S}} \right)^{0,3}$, which was found through empirical adaptions and describes the huge increase of the pressure loss in the loading zone and
- the wetting factor B_f.

The wetting factor describes the dependency of the pressure loss from the liquid load and was found through empirical adaption of Billet and Schultes in the year of 1991 (23) as given in Equations (6) and (7).

$$B_f = \exp\left(\frac{\text{Re}_L}{200} \right) \qquad \text{(Billet and Schultes 1991 (23))} \qquad (6)$$

$$\text{Re}_L = \frac{u_L * \rho_L}{a_{geo} * \eta_L} \qquad (7)$$

In the year 1999 Billet and Schultes (28) introduced a new version of the wetting factor, which can be seen in Equations (8) and (9), as this one is able to describe the experimental results at high pressures and at very low liquid viscosities better.

$$B_f = \exp\left(\frac{13300}{a_{geo}^{\frac{3}{2}}} * \sqrt{Fr_L} \right) \qquad \text{(Billet und Schultes 1999 (28))} \qquad (8)$$

$$Fr_L = \frac{u_L^2 * a_{geo}}{g} \qquad (9)$$

As reality deviates from the presented theoretical assumptions and the geometrical shape of all the different packing types cannot be described by surface area and void fraction, four empirical packing specific shape constants ($C_{P,0}$, C_h, C_S, C_{Fl}) are needed for the equations of the hydraulic part of the Billet and Schultes-Model (28). These constants are given for a lot of packing types in the literature (28) or can be calculated with the help of experimental measurement results.

Modelling of the Structured Packing Hiflow Plus

The Billet and Schultes-Model is able to predict the loading and the flooding point as well as the hold-up and the dry pressure drop satisfactorily. Furthermore, the wet pressure drop prediction of the packing Hiflow Plus #2 is acceptable as depicted in Figure 5.

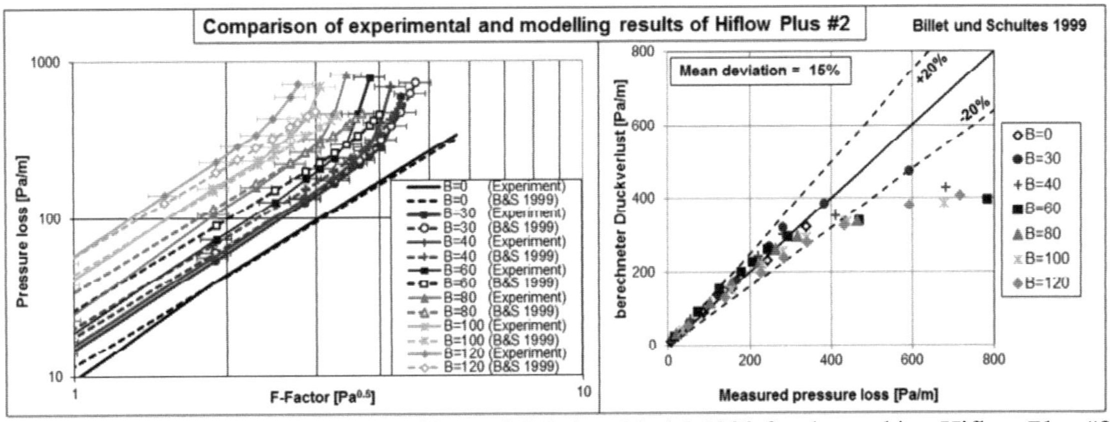

Figure 5: Modelling results by the Billet and Schultes Model 1999 for the packing Hiflow Plus #2

Nevertheless, the wet pressure drop forecast for the packing Hiflow Plus #1 does not work very well which also leads to a mean deviation over the whole considered range of 35% as can be seen in Figure 6.

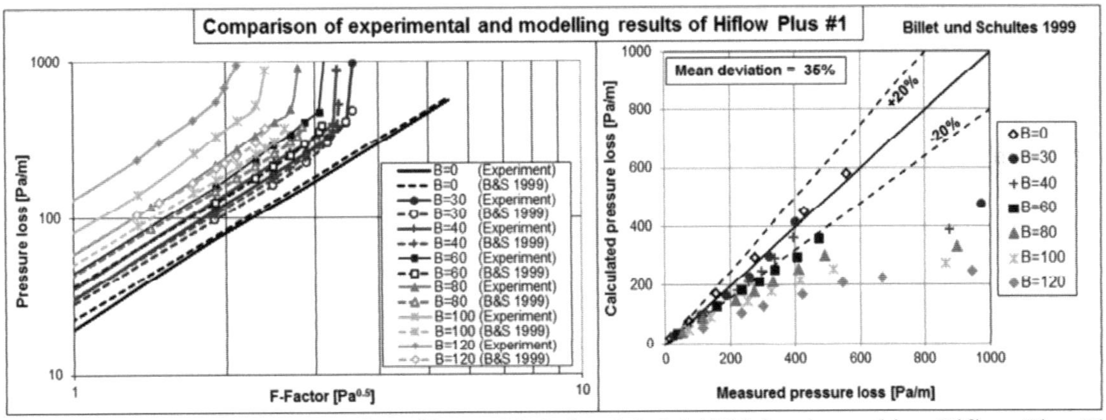

Figure 6: Modelling results by the Billet and Schultes Model 1999 for the packing Hiflow Plus #1

As the two different packings Hiflow Plus #1 and Hiflow Plus #2 have a similar geometrical structure and only differ in their packing properties (Table 1), the influence of these packing properties on the calculation of the wet pressure loss was examined as a function of gas and liquid load.

Table 1: Packing properties of Hiflow Plus

Packing	a_{geo} [m^2/m^3]	ε [m^3/m^3]
Hiflow Plus #1	180	0.93
Hiflow Plus #2	100	0.95
Hiflow Plus #3	80	0.96

In particular, the influence of the geometrical surface area on the wet pressure drop as a function of the liquid load is remarkable as shown in Figure 7, calculated by the equations given by Billet and Schultes 1991 as well as the one of 1999 for a constant gas load (F-factor = 1 $Pa^{0,5}$). It can be seen that the curve for an irrigation density of 60 $m^3/(m^2*h)$ has already a different slope than the one at an irrigation density of 30 $m^3/(m^2*h)$. At an irrigation density of 100 or 120 $m^3/(m^2*h)$ the Billet and Schultes-Model calculates a higher pressure loss for a packing with a geometric surface area of 90 m^2/m^3 as for instance for a packing with 120 m^2/m^3.

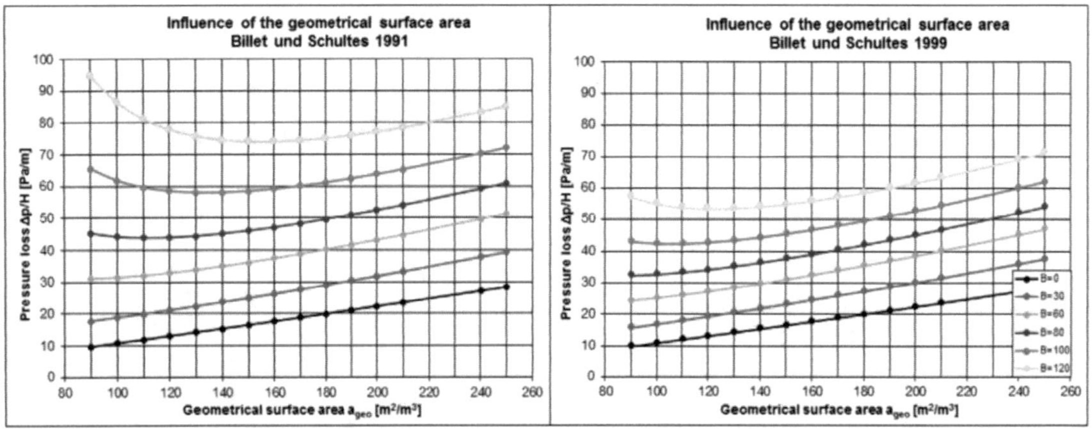

Figure 7: Influence of the geometrical surface area on the calculation of wet pressure loss

For this reason a closer consideration of the influence of the geometrical surface area a_{geo} as well as the liquid load u_L on the calculation of the wet pressure drop was accomplished in Equations (10) – (12). It shows that these two variables can both be found in the wetting factor. Furthermore, it is obvious that the influence of these two variables is similar in the two different definitions from 1991 and 1999.

$$\frac{\Delta p}{H} = \left\{ \psi_{tr} * \left(\frac{\varepsilon - h_L}{\varepsilon} \right)^{1,5} * \left(\frac{h_L}{h_{L,S}} \right)^{0,3} * B_f \right\} * \frac{a_{geo}}{(\varepsilon - h_L)^3} * \frac{F^2}{2} * \frac{1}{K} \quad (10)$$

$$B_f = \exp\left(\frac{Re_L}{200} \right) = \exp\left(\frac{1}{200} * \frac{\rho_L}{\eta_L} * \frac{u_L}{a_{geo}} \right) \quad \text{(Billet and Schultes 1991 (23))} \quad (11)$$

$$B_f = \exp\left(\frac{13300}{a_{geo}^{\frac{3}{2}}} * \sqrt{Fr_L} \right) = \exp\left(\frac{13300}{g^{0,5}} * \frac{u_L}{a_{geo}} \right) \quad \text{(Billet and Schultes 1999 (28))} \quad (12)$$

Furthermore, this relationship can be rearranged, see Equation (13): the wetting factor actually depends on the wettable length $L_{wettable}$ which is the product of the cross sectional area of the column and the specific geometrical surface area of the packing. This length gives kind of a wettable perimeter provided by the packing which the liquid can use for trickling downwards.

$$\frac{u_L}{a_{geo}} = \frac{\dot{V}_L}{A * a_{geo}} = \frac{\dot{V}_L}{L_{wettable}} \quad (13)$$

Modification of the Billet and Schultes-Model

As the wettable length used in the Billet and Schultes-Model is a constant value over the whole operation range of a column, a term which is able to describe the condition of the liquid film as a function of the packing geometry as well as of the operating conditions has to be defined. With the equations given from Billet and Schultes for the liquid hold-up as well as for the hydraulic surface area, a film thickness can be defined which depends on the Reynolds- and on the Froude-Number as described in Equation (14).

$$\delta = \frac{h_L}{a_h} = \frac{h_L = f_1(\mathrm{Re}, Fr)}{a_h = f_2(\mathrm{Re}, Fr)} \Rightarrow \delta = f(\mathrm{Re}, Fr) \tag{14}$$

By introducing this new term, and with adaption to the experimental results, a new wetting factor can be defined as shown in Equation (15). The new defined wetting factor depends on a Reynolds-Number, $\mathrm{Re}_{L,\text{film thickness}}$, which specific length is defined by the film thickness δ given in Equation (14).

$$B_f = \exp\left(\frac{1}{16} * \frac{\rho_L * u_L * \delta}{\eta_L}\right) = \exp\left(\frac{\mathrm{Re}_{L,\,film\ thickness}}{16}\right) \tag{15}$$

In Figure 8, the influence of the geometrical surface area on the calculation of the wet pressure drop is examined in the same way as shown in Figure 7. The results of the Billet and Schultes-Model 1999 and of the introduced modification of this work are compared in Figure 8. As it can be seen, the correlation of this work expresses the wet pressure loss dependence of the geometrical surface area more realistic than the Billet and Schultes-Model 1999.

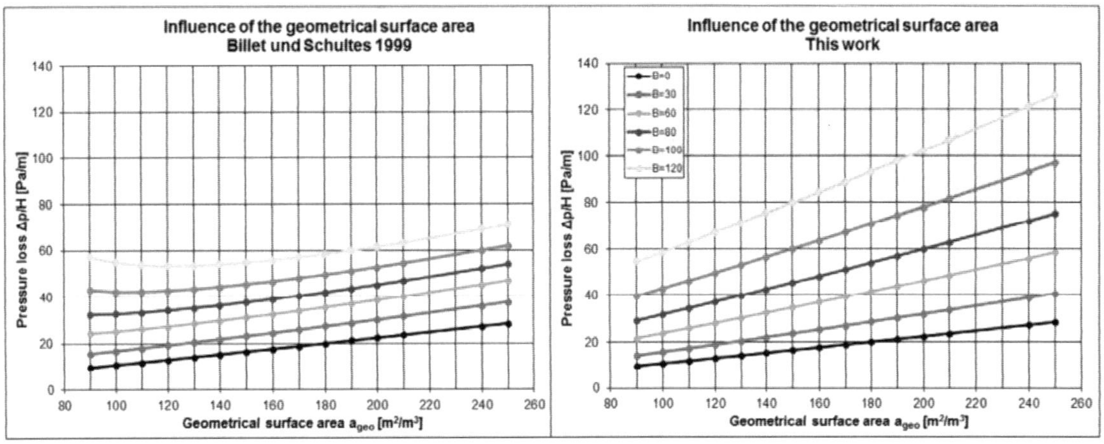

Figure 8: Influence of the geometrical surface area on the calculation of wet pressure loss

Results

In Figures 9 and 10 the experimental wet pressure loss results are compared with the calculated values of the modified Billet and Schultes-Model. It can be seen that the modelling results match better than with the original Billet and Schultes-Model, as the mean deviation for the packing Hiflow Plus #1 decreases from 35% to 15%. The deviations for the packing Hiflow Plus #2 are almost equal (15% for the Billet and Schultes-Modell 1999 and 14% for the modification).

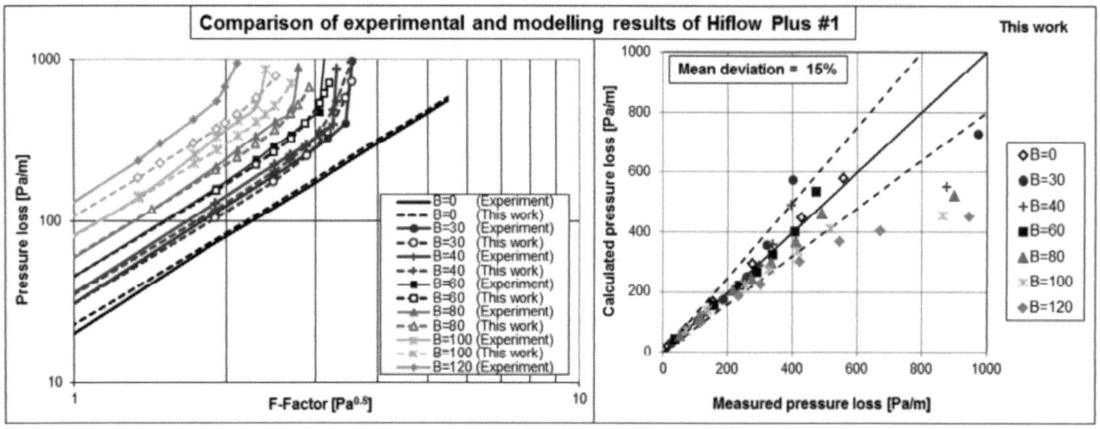

Figure 9: Modelling results by the modification of the Billet and Schultes Model for the packing Hiflow Plus #1

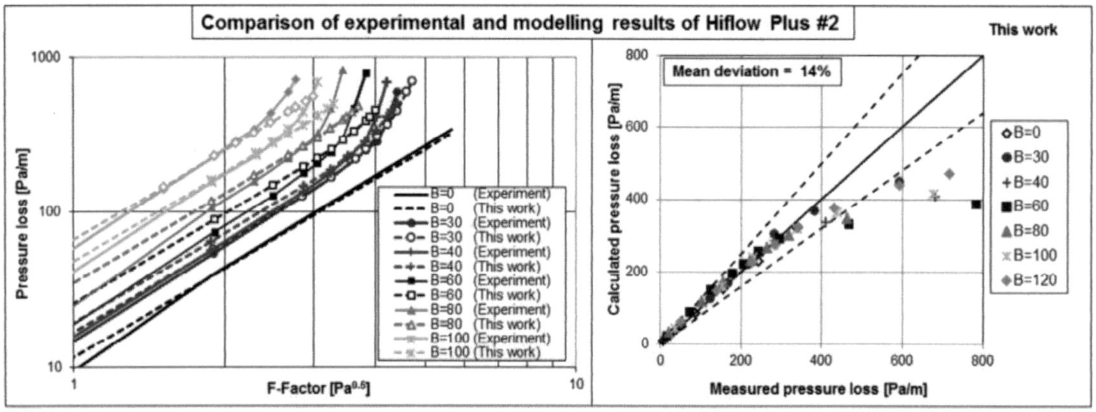

Figure 10: Modelling results by the modification of the Billet and Schultes Model for the packing Hiflow Plus #2

In order to assess the performance of this modification of the Billet and Schultes-Model, further packing types have been examined. In a first step of this validation process, two different structured plastic packings, four different random plastic packings (Table 2) as well as five different metal random packings (Table 3) have been investigated. All presented measurements were done with the system air/water under ambient conditions.

As Billet and Schultes (28) state that their wet pressure drop correlation is only valid up to an irrigation density of 60 $m^3/(m^2*h)$, the mean deviation between the measured and calculated values for this operating region was considered separately (Figure 12 and 14). Evaluating the results in the Figures 11-14, it can be summarised that the modification presented in this paper leads to significantly better forecasting results particularly for those packings which the Billet and Schultes-Model does not rate really well, as for instance for the packing RMSR 70-5. If the Billet and Schultes-Model already predicts acceptable wet pressure drops, the modification shows comparable results, e.g. packing RMSR 25-3. Furthermore, as the modification is able to predict the wet pressure drops up to an irrigation density of 120 $m^3/(m^2*h)$ with acceptable mean deviations (max. 25% for Raflux 25-5), the area of validity can be raised from an irrigation density of 0.61-60 $m^3/(m^2*h)$ up to 120 $m^3/(m^2*h)$.

Table 2: Properties of the examined plastic packing types

Packing	Material	Void Fraction [m³/m³]	Geom.Surface Area [m²/m³]
Hiflow Plus #1	PP	0.93	180
Hiflow Plus #2	PP	0.95	100
Hiflow 50-0	PP	0.94	110
Hiflow 50-6	PP	0.94	90
Hiflow 38-1	PP	0.95	150
Hiflow 90-7	PP	0.97	76

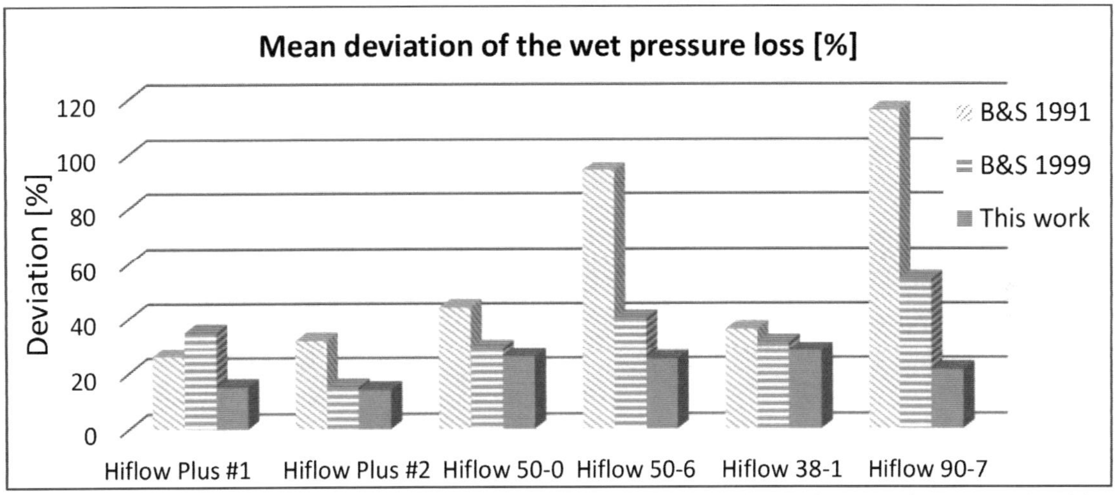

Figure 11: Mean deviation of the wet pressure drop for plastic packings for the whole examined load region

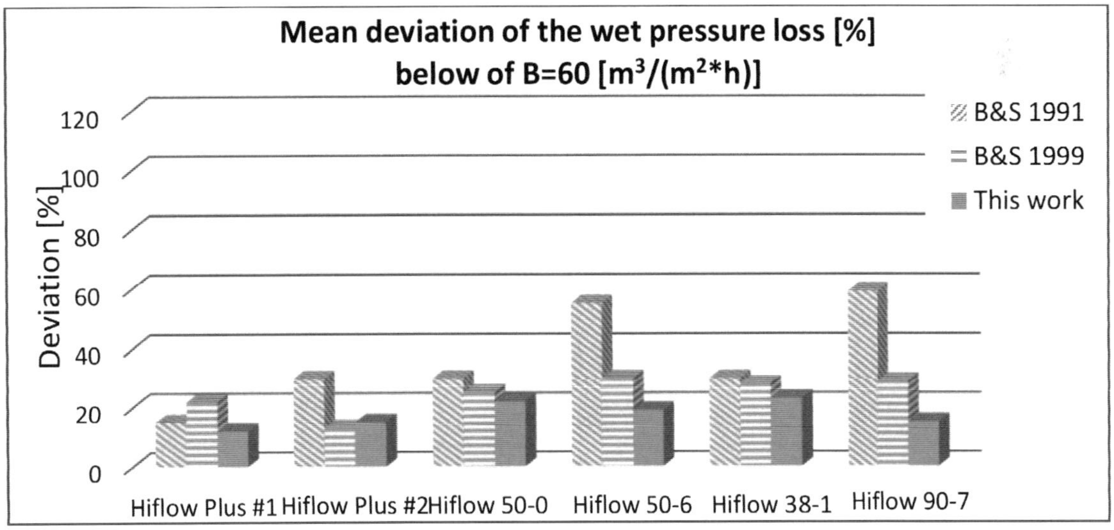

Figure 12: Mean deviation of the wet pressure drop for plastic packings considering irrigation densities only up to 60 $m^3/(m^2*h)$

Table 3: Properties of the examined metal packing types

Packing	Material	Void Fraction [m³/m³]	Geom.Surface Area [m²/m³]
Raflux 25-5	1.4404	0.95	393
Raflux 50-5	1.4306	0.97	112
RMSR 25-3	1.4000	0.97	235
RMSR 50-4	1.4301	0.98	115
RMSR 70-5	1.4000	0.98	67

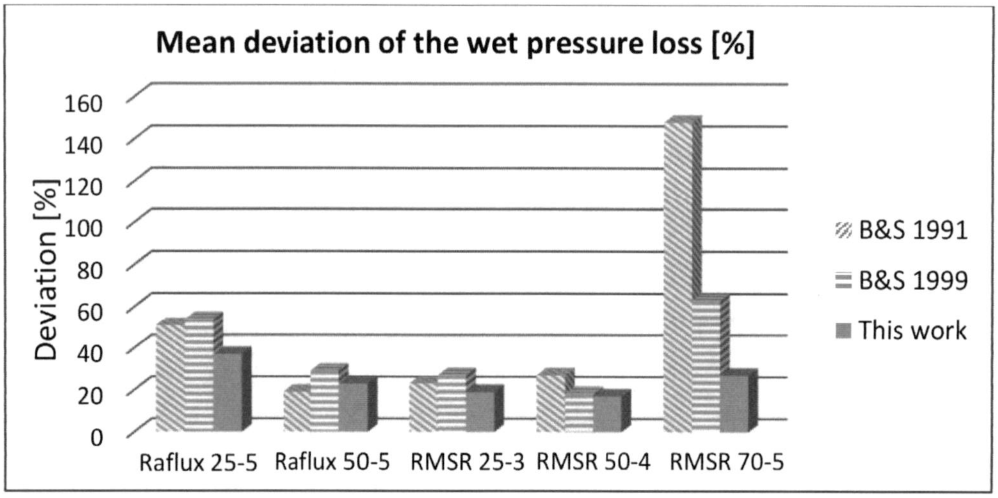

Figure 13: Mean deviation of the wet pressure drop for plastic packings for the whole examined load region

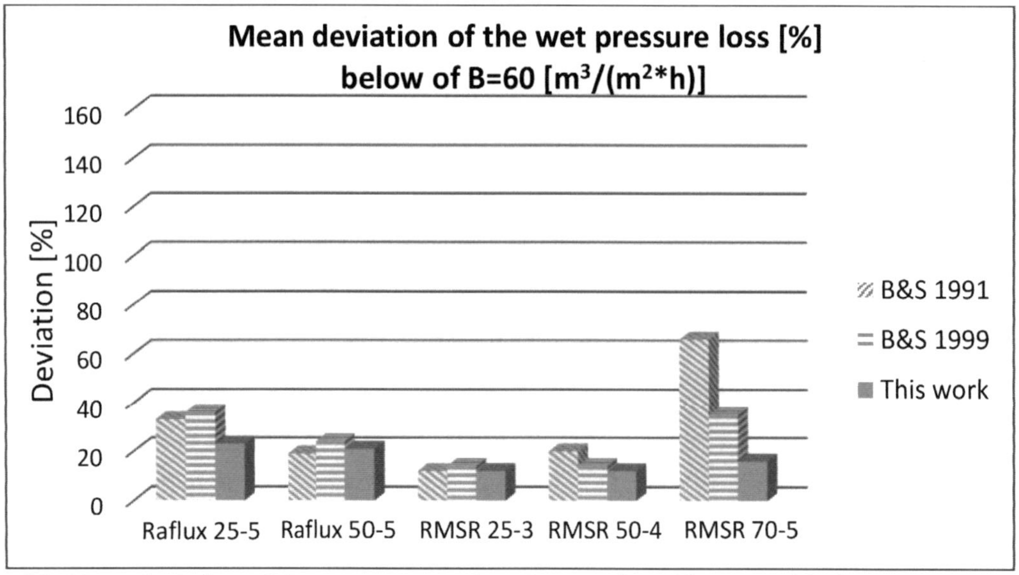

Figure 14: Mean deviation of the wet pressure drop for metal packings considering irrigation densities only up to 60 $m^3/(m^2*h)$

Conclusion and Outlook

Due to a relatively simple but physically justified modification of the Billet and Schultes-Model a better prediction of the wet pressure drop of the examined packing types (a_{geo} = 67 – 393 m^2/m^3; ε = 0.93 – 0.98 m^3/m^3) could be reached. The modification could be done by a new combination of equations given by the original model so that no new parameters have to be added. Moreover, the area of validity of the Billet and Schultes-Model of 0.61 – 60 $m^3/(m^2*h)$ can be extended to an irrigation density of 120 $m^3/(m^2*h)$ for the calculation of the wet pressure drop.

In the future further packing types, especially structured plate packings, will also be examined, and the validation will also be done with other test systems than water and air to investigate additionally the influence of the fluid properties.

Symbols

A	Column cross section [m^2]
a_{geo}	Geometrical packing surface [m^2/m^3]
a_h	Hydraulic surface area [m^2/m^3]
B_f	Wetting factor
B	Irrigation density [$m^3/(m^2*h)$]
DN	Diameter Nominal
d	Diameter [m]
d_P	Particle diameter [m] $d_P = 6 * \frac{(1-\varepsilon)}{a_{geo}}$
d_S	Column inner diameter [m]
F	Gas loading factor resp. F-factor [$Pa^{0,5}$]
f()	Function of
Fr	Froude-Number [-]
g	Gravity [N]
H	Height [m]
h_L	Hold-up [m^3/m^3]
K	Wall factor [-]
kg	Kilogram
m	Meter
Pa	Pascal
Re	Reynolds-Number [-]
s	Second
u_L	Liquid surface velocity [m/s]
$\dot{V}$	Volume flow rate [m^3/s]
<u>Greek Symbols</u>	
δ	Liquid film thickness [m]
Δ	Difference
Δp	Pressure loss [Pa]
ε	Void fraction [m^3/m^3]
η	Dynamic viscosity [kg/(m*s)], [Pa*s]
ρ	Density [kg/m^3]
Ψ	Resistance coefficient

Indices

ft	Wet
L	Liquid phase
S	At the loading point
tr	Dry

References

1. Wolf, V.; Lehner, M.; Hoffmann, K. (2012): Hydrodynamic and mass transfer processes of the grid-structured plastic packing Hiflow Plus. Filtech Exhibitions Germany (Hg.): 11. World Filtration Congress – Proceedings. Meerbusch: FILTECH, p. G14-01-059.
2. Cai, T. J.; Resetarits, M. R. (2011): Pressure drop measurements on distillation columns. Chinese Journal of Chemical Engineering 19 (5), p. 779–783.
3. Bravo, J. L.; Rocha, J. A.; Fair, J. R. (1985): Mass transfer in gauze packings. Hydrocarbon Processing 64 (1), p. 91–95.
4. Bravo, J. L.; Rocha, J. A.; Fair, J. R. (1986): Pressure drop in structured packings. Hydrocarbon Processing 56 (3), p. 45.
5. Fair, J. R.; Bravo, J. L. (1987): Prediction of mass transfer efficiencies and pressure drop for structured tower packings in vapor/liquid service. IChemE: Distillation and Absorption 87, Brighton, September 1987. London, (EFCE Publication Series 62), p. A183-A201.
6. Fair, J. R.; Bravo, J. L. (1990): Distillation columns containing structured packing. Chemical Engineering Progress 86 (1), p. 19–29.
7. Bravo, J. L.; Fair, J. R.; Rocha, J. A. (1992): A comprehensive model for the performance of columns containing structured packing. IChemE: Distillation and Absorption 1992, 2 (additional papers). Birmingham, September 1992. Rugby: (IChemE Symposium Series 128), p. A489-A507.
8. Rocha, J. Antonio; Bravo, Jose L.; Fair, James R. (1993): Distillation columns containing structured packings: a comprehensive model for their performance. 1. Hydraulic models. Ind. Eng. Chem. Res. 32 (4), p. 641–651.
9. Gualito, J. J.; Cerino, F. J.; Cardenas, J. C.; Rocha, J. A. (1997): Design method for distillation columns filled with metallic, ceramic, or plastic structured packings. Ind. Eng. Chem. Res. 36 (5), p. 1747–1757.
10. Verschoof, H.-J.; Olujić, Ž.; Fair, J. R. (1999): A general correlation for predicting the loading point of corrugated sheet structured packings. Ind. Eng. Chem. Res. 38 (10), p. 3663–3669.
11. Olujić, Ž. (1997): Development of a complete simulation model for predicting the hydraulic and seperation performance of distillation columns equipped with structured packings. Chemical and Biochemical Engineering Quaterly 11 (1), p. 31–46.
12. Olujić, Ž.; Behrens, M.; Spiegel, L. (2007): Experimental characterization and modeling of the performance of a large-specific-area high-capacity structured packing. Ind. Eng. Chem. Res. 46 (3), p. 883–893.
13. Piché, S.; Iliuta, I.; Grandjean, B. P.A.; Larachi, F. (2001a): A unified approach to the hydraulics and mass transfer in randomly packed towers. Chemical Engineering Science 56 (21-22), p. 6003–6013.
14. Piché, S.; Larachi, F.; Grandjean, B.P.A. (2001b): Improved liquid hold-up correlation for randomly packed towers. Chemical Engineering Research and Design 79 (1), p. 71–80.
15. Piché, S.; Larachi, F.; Grandjean, B.P.A. (2001c): Loading capacity in packed towers – database, correlations and analysis. Chemical Engineering & Technology 24 (4), p. 373–380.

16. Piché, S.; Larachi, F.; Grandjean, B. P. A. (2001d): Flooding capacity in packed towers: database, correlations and analysis. Ind. Eng. Chem. Res. 40 (1), p. 476–487.
17. Piché, S. R.; Larachi, F.; Grandjean, B. P. A. (2001e): Improving the prediction of irrigated pressure drop in packed absorption towers. Can. J. Chem. Eng. 79 (4), p. 584–594.
18. Ortiz-Arroyo, A.; Larachi, F.; Iliuta, I. (2003): Method for inferring contact angle and for correlating static liquid hold-up in packed beds. Chemical Engineering Science 58 (13), p. 2835–2855.
19. Billet, R. (1987): Modelling of fluiddynamics in packed columns. IChemE: Distillation and Absorption 87, Bd. 1. Brighton, 07.-09. September 1987. London, (EFCE Publication Series 62), p. A171-A182.
20. Billet, R. (1988): Relationship between residence time, fluid dynamics and efficiency in countercurrent flow equipment. Chem. Eng. Technol. 11 (1), p. 139–148.
21. Billet, R.; Schultes, M. (1987a): Capacity studies on two-phase countercurrent-flow columns. IChemE: Distillation and Absorption 87, Bd. 2. Brighton, 07.-09. September 1987. London, (EFCE Publication Series 62), p. B255-B266.
22. Billet, R.; Schultes, M. (1987b): Determination of liquid hold-up in gas-liquid two-phase countercurrent mass transfer columns. IChemE: Distillation and Absorption 87, Bd. 1. Brighton, 07.-09. September 1987. London, (EFCE Publication Series 62), p. A159-A170.
23. Billet, R.; Schultes, M. (1991): Modelling of pressure drop in packed columns. Chem. Eng. Technol. 14 (2), p. 89–95.
24. Billet, R.; Schultes, M. (1992): Advantage in correlating packed column performance. In: IChemE: Distillation and Absorption 1992, Bd. 1. Birmingham, 07.-09. September 1992. European Federation of Chemical Engineering. Rugby, (IChemE Symposium Series 128), p. B129-B136.
25. Billet, R.; Schultes, M. (1993): A physical model for the prediction of liquid hold-up in two-phase countercurrent columns. Chem. Eng. Technol. 16 (6), p. 370–375.
26. Billet, R. (1995): Packed towers in processing and environmental technology. 1. ed. Weinheim, VCH.
27. Billet, R.; Schultes, M. (1995): Fluid dynamics and mass transfer in the total capacity range of packed columns up to the flood point. Chem. Eng. Technol. 18 (6), p. 371–379.
28. Billet, R.; Schultes, M. (1999): Prediction of mass transfer columns with dumped and arranged packings. Chemical Engineering Research and Design 77 (6), p. 498–504.
29. Maćkowiak, J. (1990): Bestimmung des Druckverlustes berieselter Füllkörperschüttungen und Packungen. Staub - Reinhaltung der Luft 50, p. 455–463. (german)
30. Maćkowiak, J. (2003): Fluiddynamik von Füllkörpern und Packungen. Grundlagen der Kolonnenauslegung. 2nd edition Berlin, Springer. (german)
31. Maćkowiak, J. (2006): Neue Aspekte der Modellierung bei der Fluiddynamik von Packungskolonnen. Chemie Ingenieur Technik 78 (8), p. 1079–1086. (german)
32. Stichlmair, J.; Bravo, J. L.; Fair, J. R. (1989): General model for prediction of pressure drop and capacity of countercurrent gas/liquid packed columns. Gas Separation & Purification 3 (1), p. 19–28.
33. Engel, V. (1999): Fluiddynamik in Packungskolonnen für Gas-Flüssig-Systeme. PhD Thesis. TU München, München. Chair of Fluiddynamics. (german)
34. Engel, V.; Stichlmair, J.; Geipel, W. (2001): Fluid Dynamics of Packings for Gas-Liquid Contactors. Chemical Engineering Technology 24 (5), p. 459–462.
35. Ergun, S. (1952): Fluid flow through packed columns. Chemical Engineering Progress 48 (2), p. 89–94.

402171 How Sensitive Is Your Treating Plant to Operating Changes – Part 2

Abstract:

One of the most neglected areas in gas treating is the proper selection of tower internals in both absorbers and regenerators. Apart from hydraulic considerations, the mass transfer performance of tower internals unfortunately has tended to be almost completely ignored. When parts-per-million specifications must be met on product gases, selecting the wrong packing size or bed depth can result in a failed design. For many years, packing has had a bad reputation in absorption and distillation at high pressure. Part of the reason is inattention to proper distributor design. Another is the persistent and still unresolved difficulty in translating ideal stages to actual packed bed depths and the selection of a particular commercial packing. The effect in the contactor's diameter and height of different random packing sizes and types is compared to structured packing and two types of commercially proven valve trays for an LNG project under study that initially will process 745 KNCMH of a gas containing 11.5% of CO2.

401530 Experimental Investigation of the Effects of Tower Motion on Random Packing Efficiency

Abstract:

Efforts to monetize offshore gas resources have led to increased interest in floating processing facilities such as Floating LNG. These facilities face technical challenges, including the proper design of process equipment to achieve desired performance under moving conditions. In particular, tilt and motions imposed on a fractionation column can have a significant impact on product specifications due to reduced packing efficiency. An improved understanding of motion impacts will assist the analysis of tower performance and enable fit-for-purpose designs to deliver reliable performance and minimize overall cost of floating production facilities.

The current understanding of towers in motion allows only very basic design rules and the safety factor for packing height in literature varies widely. Therefore, an experimental test program was conducted with a 1-m diameter test column to quantify packing hydraulics and mass transfer performance with simulated wave motion. The test column contained random packing as it is typically preferred in amine gas treating and high pressure distillation towers for gas processing plants. The mass transfer efficiency was measured with carbon dioxide absorption into caustic solution to obtain the effective wetted packing area for mass transfer.

This paper will present experimental results for the impact of design and operating parameters (such as packed bed height, tilt angle, etc.) on the mass transfer performance of a modern random packing. These results are also analyzed in terms of dry/under-wetted packing volume and vapor channeling in order to obtain a model that may be used for packed tower design.

The design of floating processing facilities requires a good understanding of the impact of motion on fractionation tower performance. Unlike previous studies, which have largely been limited to hydraulic measurements of liquid maldistribution, theoretical predictions of performance, and/or small scale experiments, this paper will present mass transfer measurements with a modern, industrial scale random packing. The improved understanding of packing performance under motion will facilitate tower designs that meet process functional requirements and minimize overall cost of floating production facilities.

400959 Packing Characterization for Post Combustion Capture: Model Development

Abstract:

Packing is widely used in post-combustion CO_2 capture because of its low pressure drop, good mass transfer efficiency, and ease of installation. In the CO_2 capture process, absorber and stripper performance are highly dependent on the effective mass transfer area of the packing (a_e). The stripper performance also depends on the liquid film mass transfer coefficient (k_L). Gas cooler and water wash performance depends on the gas film mass transfer coefficient (k_G). In this work, three mass transfer models predicting a_e, k_L, and $k_G$have been developed based on measurements in the 0.428 m diameter PVC column, including the data for eleven structured packings and three random packings.

The effective mass transfer area model is based on previous researcher's model (Tsai, 2010). The effective area is assumed to be only a function of liquid flow rate, packing surface area, surface tension, and is independent of gas flow rate and liquid phase viscosity. In this work, the improvement from Tsai's model is to use liquid superficial velocity over packing total area (u_L/a_P) as the liquid flow rate per perimeter. Therefore, the area model can be applied in situations where channel dimensions are not known or hardly defined. The effective area model is:

$$a_e/a_P = 1.41[(\rho_L/\sigma)g^{1/3}(u_L/a_P)^{4/3}]^{0.116}$$

The k_L and k_G models are based on three factors that influence packing mass transfer: liquid or gas superficial velocity ($u_{L/G}$), the packing surface area (a_P), and the mixing point density (M). Mixing point density is a new concept proposed in this work, which is the number of contacting points between packing corrugated metal sheets. It represents the packing geometry effect on mass transfer coefficients. Mixing point density can be calculated by packing channel base B, crimp height h or packing surface area $a_P$and corrugation angle in a more general case:

$$M = 6/(BhB*\tan\theta) = (3a_P^3\sin\theta\cos\theta)/[16(\sin^2\theta+1)^{3/2}]$$

The k_L and k_G models devloped in this work are:

$$k_L=3.08E\text{-}3u_L^{0.72}M^{0.42}a_P^{-1.15}$$

$$k_G=9.6E\text{-}3u_G^{0.54}M^{0.29}a_P^{-0.5}$$

The mass transfer models developed in this work can predict experimental data well. The average deviation is 10% for effective area model, 22% for k_L model, and 13% for k_G model. The models are also compared with literature models and have good comparison.

402169 High Pressure Distillation with Structured Packing - Friend or Foe

Abstract:

Through the 1980's and 90's, the industry reported catastrophic failures when structured packings were employed in high pressure distillation. In 2001 Andrew Sloley, of The Distillation Group, set out criteria where structured packing should be avoided in high pressure distillation. The purpose of this paper is to quantify different physical properties where structured packing is most and least favored. Data were taken from FRI published Mellapak 250.Y data at eight different operating pressure. It will be shown that low surface tensions and high vapor densities are major indicators where structured packings should be avoided.

402170 Are Stacked Packings Structured Packings? Fluid Dynamics and Their Separation Efficiency for Gas-Liquid Systems

Abstract:

Random packing elements can be regularly arranged inside the column to form a regular structure as so-called stacked packings. This filling method of random packings is not new, it is known for quite some time in the separation techniques. For instance, hydraulics and mass transfer of stacked 50-150 mm ceramic Raschig rings have been investigated from 1960-1975 in the context of absorption processes for sulphuric acid processing [1-3].

For stacked packings consisting of Pall rings, Bia³ecki-rings or even modern lattice type packings such as Hiflow-ring and Nor-Pac no systematic investigations and no general approach for designing mass transfer processes for gas-liquid systems with stacked packings are present in literature.

Very good results were obtained for stacked 25-50 mm metallic Bialecki-rings in the liquid-liquid extraction processes, which werde presented in the literature [4].

Generally it is known, that stacked packings have a bigger operating range compared to random packings and a much lower pressure drop on one hand, but lower separation efficiency on the other.

Thus, which advantages have stacked packings? Can they be used as an alternative to random or to structured packings? And finally: Should stacked packings be regarded as structured packings or rather as random packings?

The main goal of this work is to provide the answers for the questions raised above based on detailed and systematic investigations. Another aim is to present their range of application for which they can be recommended as an alternative column filling to random and structured packings.

In this paper , a generally valid model for description of fluid dynamics [5] and mass transfer in the gas- and liquid phase of different stacked packing types such as 50 mm Raschig rings, 50 mm Pall rings, 25-50 mm Bialecki rings, 50 mm Nor Pac and 50 mm Hiflow rings made of ceramics, metal and plastics will be presented. The model is based on experimental data for different absorption and distillation systems such as NH3-air/water, NH3-water/air, CO2-water/air under normal pressure, chlorobenzene-ethylbenzene, ethanol-water in the pressure range from 13 mbar to 1000 mbar. Experiments have been carried out in columns with diameters ds= 0,15 – 0,6 m and bed heights up to 2m.

After that a comparison of stacked packing and structured packing and random packing is carried out and the obtained results are discussed.

As main results of this work it can be concluded, that the behaviour of stacked packing is similar to that of structured packing and processes are discussed in which this type of packed column can be applied successfully.

Literature

[1] W.M. Ramm

Absorption of gases (orig. Russian), Moscow (1966)

[2] Z. Zio³kowski, J. Maækowiak, J. Kowalski

Liquid phase mass transfer in packed column filled with stacked Bia³ecki rings (orig. polish)

Inz. Chem. (1975), No 3, p.631- 640

[3] J. Maækowiak, S. Suder

Hydraulics and mass transfer behaviour in packed column filled with stacked Bia³ecki-Rings for gas/liquid-systems (orig. polish) Inz. Chem. (1977) Vol. VII, No.3, p.651 – 664

[4] J. Maækowiak, R.Billet

New Method for Design of Packed Columns with Random and Stacked Packings for Liquid-Liquid-Extraction (orig. englisch)

Ger. Chem. Eng Vol.1 (1986), p.48 - 64

[5] J. Maækowiak

Fluid Dynamics of Packed Columns

Springer-Heidelberg , London, New York 2010

A STUDY OF AXIAL MIXING INFLUENCE ON THE MASS TRANSFER TRANSFERABILITY FROM ABSORPTION TO DISTILLATION CONDITIONS

Lukas Valenz, Frantisek J. Rejl[1], Jan Haidl
University of Chemistry and Technology, Prague, Czech Republic

Introduction

Distillation and absorption are treated as analogical mass-transfer processes between liquid and gas phase on the packing. The analogy between processes assumes that the same set of models can be used for description of both processes, while there exists a clear relation between their parameters. While description of both processes by complete set of the models is theoretically possible, practically it faces insurmountable difficulties due to lack of the model parameters and computational complexity. Therefore deep simplification is often introduced and reflects the specifics of the individual process. Among these simplifications belongs description of the phase hydrodynamics by a plug flow model which doesn't need any parameter for its utilization.

In our previous contribution at AIChE Spring meeting 2014 we have presented the HETP data of RSP250Y packing recalculated from the k_La and k_Ga data obtained by standard absorption experiments using Billet-Schultes model. The *HETPs* calculated out of the simulations were barely half of the experimental ones when the plug-flow hydrodynamic model was used. On the other hand, when axial dispersion in the liquid phase was taken into account, the out of simulation calculated *HETP*s matched the experimental ones with only about 16 % and 9 % error (for C_6/C_7 and MeOH/PrOH system, respectively). The acquisition of the absorption mass-transfer data and their recalculation from absorption to the distillation conditions seemed to be therefore successful in this particular case, but incorporation of the more sophisticated hydrodynamic model was clearly inevitable.

In this paper we present result of similar procedure for larger spectra of packings and systems and also present axial mixing data of the packings evaluated from the distillation experiments. Analysis of the distillation data revealed also relative impact of the individual phenomena (mass-transfer resistance in the liquid/gas phase, axial mixing of the liquid phase) on the separation efficiency.

Methods

Mass-transfer coefficients and effective interfacial area of the packings under absorption conditions has been measured using standard aqueous systems (O_2/water – k_La, SO_2/aq. NaOH - k_Ga, CO_2/aq. NaOH - a_{EFF}) for various flows of the gas and liquid phase. These data has been used for calculation of the HETP; the Billet-Schultes correlation was used for recalculation of the data to the distillation conditions. The axial mixing in the liquid phase has been measured using dynamic experiments involving NaCl tracer with the air/water system. The axial mixing in the gas phase has been measured

[1] presenting and corresponding author

Paper to be presented at AIChE Spring meeting, Austin, TX, 26th – 30th April 2015 using helium tracer with the same system and was found negligible with respect to that in the liquid phase. Experiments have been performed in the i.d. 0.295m column.

Distillation experiments have been performed with MeOH/EtOH and C_6/C_7 systems in the atmospheric distillation column of i.d. 0.15m under total reflux. *HETP* has been evaluated out of the composition in the head and in the bottom. Molar fraction profiles in the liquid and in the gas phase have been measured and were used for simulation of the column behavior out of which k_La, k_Ga and axial mixing (in terms of Bo_L) have been evaluated.

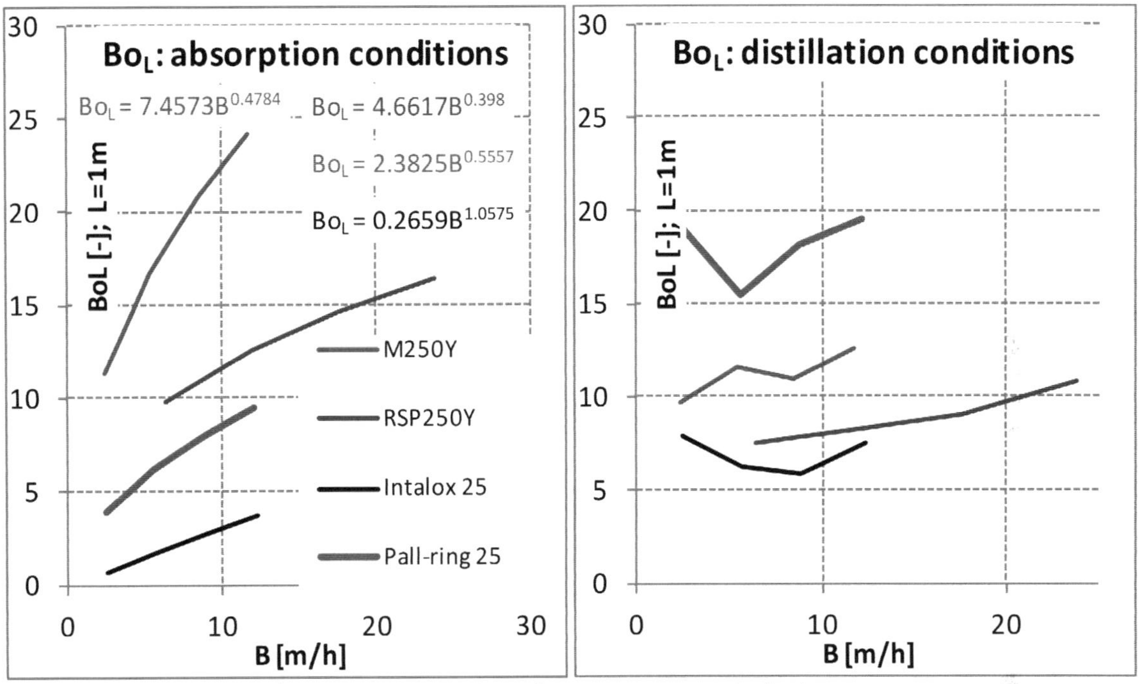

Fig. 1A,1B: axial mixing AIR/WATER system (1A), axial mixing DISTILLATION (1B)

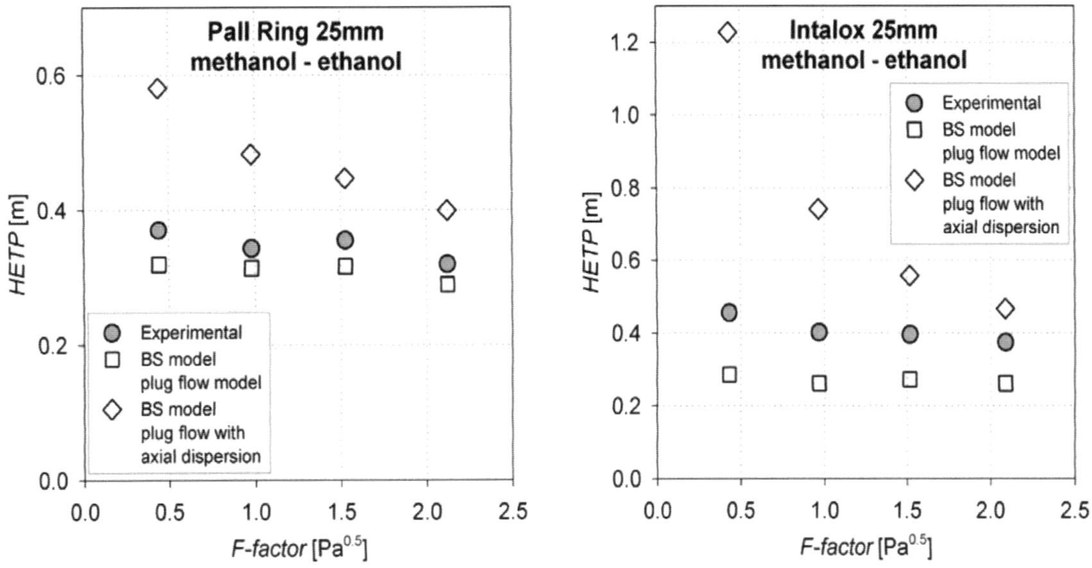

Fig. 2A

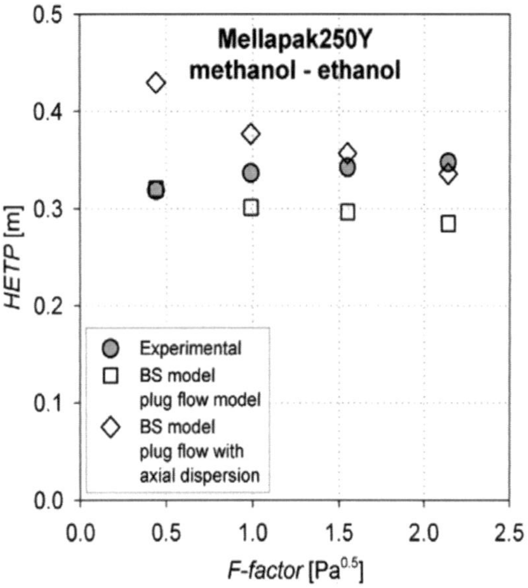

Fig. 2B

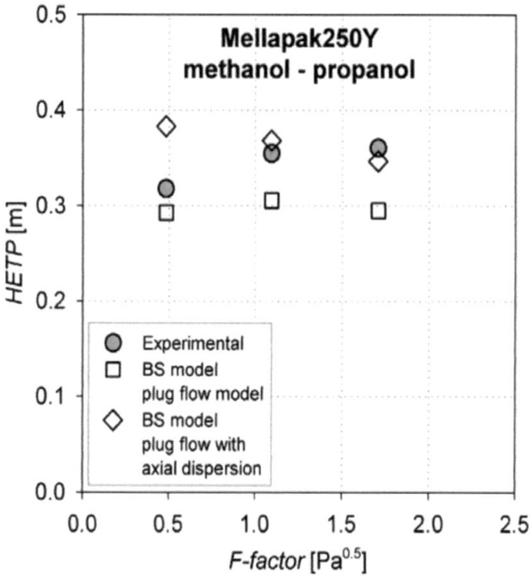

Fig. 2C

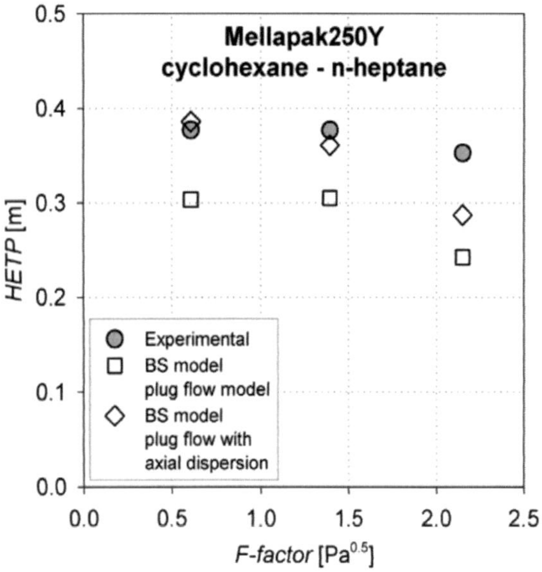

Fig. 2D

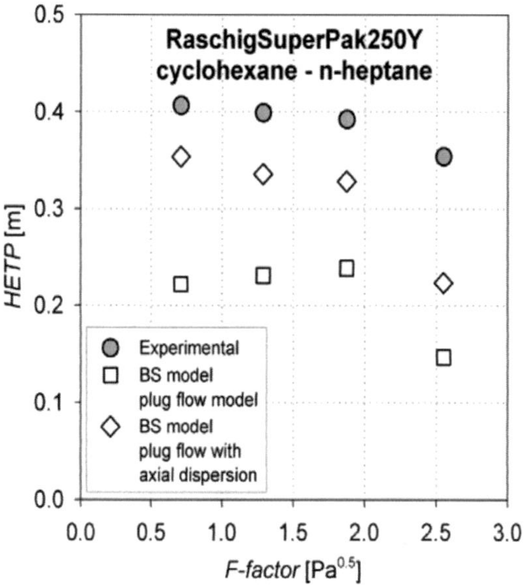

Fig. 2E

Fig. 2F

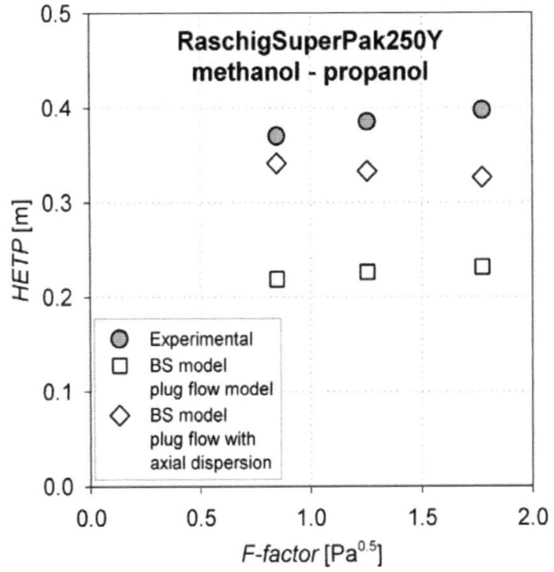

Fig. 2G

			RELATIVE EFFECT OF INDIVIDUAL PHENOMENA ON SEPARATION EFFICIENCY (UNDER DISTILLATION CONDITIONS)		
PACKING	SYSTEM	LOAD	MASS TRANSFER IN THE LIQUID PHASE	MASS TRANSFER IN THE GAS PHASE	AXIAL MIXING OF THE LIQUID PHASE
M250Y	MeOH/EtOH	10kW	18%	50%	32%
M250Y	MeOH/EtOH	40kW	16%	63%	21%
M250Y	C_6/C_7	8kW	18%	67%	15%
M250Y	C_6/C_7	24kW	26%	62%	13%
Pall 25	MeOH/EtOH	10kW	15%	72%	13%
Pall 25	MeOH/EtOH	40kW	23%	62%	15%
Intalox-25	MeOH/EtOH	10kW	13%	64%	23%
Intalox-25	MeOH/EtOH	40kW	8%	64%	28%
RSP250Y	C_6/C_7	9kW	18%	58%	24%
RSP250Y	C_6/C_7	28kW	24%	53%	24%

Tab. 1: Relative effect of individual phenomena on separation efficiency (distillation conditions)

Results

On the basis of the experiments performed in the absorption column the axial mixing is found to occur in the largest scale for the Intalox-25 packing (low Bo_L values), to be of medium scale for Pall-ring 25 and Raschig SuperPak 250Y and the smallest was found for Mellapak 250Y (high Bo_L values) (see Fig 1A). Generally the axial mixing was found to be higher for random packings than for structured packing and the values of Bo_L were found to increase with the flow of the liquid phase (with the power of 0,4-1,06).

HETP calculated out of the absorption k_La, k_Ga is for all packings and systems too optimistic (too low) with respect to the experimental ones (see Fig 2A-2G). The HETP calculated using air/water axial mixing data (recalculated to the distillation conditions with Fairs correlation) overestimates the distillation *HETP* data significantly for random packings while for the RSP250Y introduction of the axial mixing improves the agreement. For the M250Y packing the results are inconclusive (see Fig 2A-2G).

The axial mixing data evaluated from the distillation experiments (see Fig 1B) confirm the highest degree of axial mixing for the Intalox-25 packing. However, severe mixing is also found for Raschig SuperPak 250Y and axial mixing on the Mellapak 250Y seems to be much more intense under distillation conditions than for the air/water system. On the other hand the mixing on Pall Ring 25 is found lower than for the air/water system and even lower than that for Mellapak 250Y. The strong dependence of the Bo_L on the liquid phase flow vanishes under distillation conditions.

Assessment of the relative importance of the individual phenomena not surprisingly attributes major influence to the gas-side mass-transfer resistance (see tab. 1), which forms at least 50% of the "resistance against the separation" for all packings and systems. However, while for the Pall rings and for MeOH/PrOH distillation on M250Y the second most important factor is the liquid-side mass-transfer resistance, for other cases it is the axial mixing of the liquid phase.

Conclusions

Present modeling of the absorption and distillation process often assumes plug flow of the phases in the column. This assumption was found unjust, as the axial mixing forms 13-32% of the overall resistance against separation and is in fact one of the most important phenomenon which occurs in these processes. Relatively poor *HETP* results of the Intalox-25 are not caused by some extremely low values of k_La and k_Ga, but by severe axial mixing in the liquid phase on this packing. This finding should be considered also from the point of view of a new packing design.

Usage of the plug-flow model for evaluation of the k_La, k_Ga data out of standard absorption experiments leads to involution of the mixing effect in their values. The same holds for the distillation experiments, but the effect unfortunately doesn't cancel. The effect of axial mixing on the evaluated k_La, k_Ga values is different for both processes as: 1) the axial mixing in terms of the Bo_L number might differ considerably under absorption and distillation conditions 2) the effect of axial mixing on the overall separation efficiency is strongly dependent on the number of transfer units – in which the standard absorption and distillation experiments usually differ also.

The axial mixing in the liquid phase doesn't have to be only phenomenon whose neglect causes difficulties in common description of the absorption and distillation process. It was shown, however, that without taking it into account the successful transfer of the data from one process to another is achieved only by chance.

Acknowledgement

Authors are grateful to Raschig the Czech Science Foundation for supporting the project through Grant 13-01251S, and to the financial support from specific university research (MSMT No 20/2015).

References

1. Billet, R. and Schultes, M. (1999), „Prediction of Mass Transfer Columns with Dumped and Arranged Packings. Updates Summary of the Calculation Method of Billet and Schultes.", in *Trans IChemE*, 77 (Part A), p. 498-504.

Symbols

a_{EFF}	effective interfacial area of the column, $m^2 \cdot m^{-3}$
k_Ga	volumetric gas phase mass-transfer coefficient, s^{-1}

$k_L a$	volumetric liquid phase mass-transfer coefficient, s^{-1}
HETP	height equivalent to the theoretical plate, m

Subscripts

G	gas phase
L	liquid phase

400153 Monitor Fouling in Distillation Towers with Gamma Scanning

Abstract:

Henry Kister reports that plugging or fouling is the Number 1 malfunction in distillation/separation towers[1]. There are several types of fouling mechanisms though coking is more prevalent in refinery operations while scale formation or polymer fouling prevails in chemical plants.

In the case of any type of fouling inside a tower gamma scans can be very effective at pinpointing the location and severity of the problem in both trayed and packed towers.

Baseline scans while the tower is clean provides a benchmark that serves as a reference for comparing with future fouling conditions. Subsequent scans become much more meaningful for diagnosing and monitoring the fouling accumulations. In the last few years some plant operators have begun to proactively monitor and mitigate their fouling problems by using gamma scans to routinely measure the presence and extent of this problem.

Three case studies are presented showing advantages of monitoring fouling tendencies with gamma scans. These include a trayed tower showing where flooding starts due to fouling and how it progresses up the tower over time, a packed tower showing how fouling in the bottom of a packed bed builds and increases liquid retention over time and tomography monitoring for coke buildup in a refinery vacuum tower packed bed.

[1] Kister, Henry Z., Distillation Troubleshooting, Hoboken NJ, J. Wiley & Sons, Inc., 2006.

402175 Understanding Tray Flooding Phenomena

Abstract:

Tray capacity and efficiency limitations are problematic for companies operating distillation towers. When troubleshooting a distillation tower the expectation is normally that the mechanism causing the limitation would be identified and that it could be addressed in a revamp. For this to happen it is important that the troubleshooter be able to distinguish between the mechanisms that cause these limitations. In this paper it will be shown how plant data and tower scans can be used to identify the mechanism causing the tower limitation

402177 Quick and Easy Troubleshooting of a Packed Column: Thermal Imaging As a Novel Method

Abstract:

This paper presents thermal imaging as a successful troubleshooting method for packed distillation towers. A distillation tower showed poor performance in the top section, indicated by an increased and unstable top temperature. The methodology of investigation combined surveys drawn from the equipment's available instrumentation on temperature, pressure and flow, together with taking thermal images to identify possible root causes for the poor performance. The thermal images explicitly uncovered severe maldistribution throughout the tower, ranging from the top packed bed and pumparound system all the way down the bottom of the tower. The thermal images clearly show the vapour channeling up one side of the column with the liquid channeling down the other side; there were no indications of flooding. A damaged holddown grid, pumparound liquid distributor and random packing were identified as the main possible root causes for the poor performance prior to entering the column. The column internals were inspected and it was found that the random packing in the top bed was fouled together in conglomerates and severely crushed. Some of the random packing conglomerates was stuck in the holddown grids, possibly contributing to the maldistrubution. The maldistribution was completely rectified with the installation of a new bed limiter together with new random packing. This study shows that thermal imaging can effectively be used to troubleshoot packed columns on an industrial scale.

398606 Steps for Troubleshooting Amine Sweetening Plants

Abstract:

Meeting: Spring AIChE Meeting

Austin, TX – April 26th through April 30th, 2015

Distillation 2015

Title: *"Steps for Troubleshooting Amine Sweetening Plants"*

Authors: Glenn Shiveler, Sulzer Chemtech USA Inc. – Tulsa, OK

Harold Wandke, Sulzer Chemtech USA Inc. – Tulsa, OK

Abstract:

Amine gas treating (also called "sweetening") is a well-proven method for removing acid gases from a gas stream, and is commonly used in natural gas plants, petrochemical plants, and refineries. An amine unit typically consists of two towers—a contactor to allow the selected amine to remove the acid gases, followed by another column that regenerates the amine using a reboiler or direct steam injection, plus the associated heat exchangers.

The proper design and operation of the plant is critical to meet product quality and environmental standards while minimizing operating costs to the plant. There are multiple potential problems that can result in an underperforming amine sweetening plant, including inadequate hydraulic capacity of the equipment, damaged equipment, non-optimal operating conditions, heat stable salts, and poor amine quality. Determining the cause of a malfunctioning sweetening plant can be difficult because the investigating engineer can either misdiagnose the root cause of the problem, or fail to identify all of the contributing factors to the primary problem. Deductive reasoning is a critical element in the review of the potential causes of the problem.

This paper presents a list of recommended steps for performing a thorough investigation of an underperforming amine sweetening plant. The steps include collecting operating plant data to eliminate unlikely causes and then focusing in on the probable causes based on the plant data. A discussion of the role of thermal scan and gamma-scan data to diagnose hydraulic problems with the mass transfer equipment is included. Finally, guidelines for key operational parameters for the safe operation of tray hydraulics and packing capacity will be presented.

399060 Stripper Tower Capacity Problem Resolved Using Gamma Scanning Technology

Abstract:

Process Engineers have a wide array of tools to investigate operational problems. These can range from studying process variables such as temperature, pressure, flow rates, and analytical results to more advanced modeling and simulation. On many occasions the scope or priority of the troubleshooting effort is governed by a critical situation and the potential seriousness of a problem requiring quick action and parallel efforts.

Gamma scanning has gained in popularity over the years because it quickly provides process information that is not available from any other source, is real time, and non-intrusive. The technology is flexible and can be used at the onset of a problem to diagnose internal hydraulic issues, used later in an investigation to support other operating parameters pointing to where a problem may lie or validate a hypothesis related to what a particular problem may be.

This presentation offers a troubleshooting case study concerning a waste water stripper tower. A gamma scan was used to help find the root cause of a loss in capacity prior to shut down and entry. The gamma scan determined where problems existed and allowed focus during shutdown resulting in a shorter outage.

This presentation includes system description, explanation of trouble-shooting techniques used, gamma scan results and pictures from inside the tower during the outage inspection corresponding to features detected by the gamma scan.

402178 Flooded Condensers Controls: Principles and Troubleshooting

Abstract:

Flooded condenser control is most popular with total condensers generating a liquid product only. The principles of flooded condenser control methods had been described in the literature several times over the last 60 years, yet some of these methods, like the hot vapor bypass and the flooded drum methods, continue to be among the most troublesome distillation controls in chemical plants and refineries. Most of the reported problems could have been avoided with a better understanding of the principles and learning from past experiences. This paper provides a detailed description of the principles and applies them to address many of the most common traps that had caused problems in the application of these methods.

With flooded condenser controls, some of the condenser surface is flooded with liquid at all times. The flow of condensate from the condenser is directly or indirectly manipulated to vary the flooded area. Little heat transfer takes place in the flooded regions, as heat is transferred by sensible heat exchange only. To raise column pressure, the flow of condensate from the condenser is lessened. This increases the flooded area in the condenser and lowers the surface area exposed for vapor condensation. This in turn reduces condensation rate, thereby raising pressure.

This paper discusses the hydraulics of flooded condensers, and presents the key equations. It emphasizes the interaction between the hydraulics and the vapor pressure behavior of the reflux drum liquid. Lack of understanding of the role of vapor pressure in these methods has bred many of the operating problems with flooded condenser methods. Our paper then discusses experiences with the main operating issues: poor vapor and/or liquid line hook-up, poor decanting of water, presence of non-condensables, insufficient subcooling, the impact of heat loss in uninsulated reflux drums, and considerations unique to specific common variations of the flooded condenser method.

401315 Entrainment of Heavies from a Tray Column

Abstract:

Entrainment of Heavies from a Tray Column

Quality of Mono Ethylene Glycol (MEG) got deteriorated on April 21, 2012, for the first time since the commissioning of Di-Methyl Carbonate (DMC) train. UV 220nm transmittance of MEG reduced to almost nil, as against the specification of 80% minimum. This problem was investigated by collecting plant data, running simulation models and carrying out laboratory tests.

The decrease in UV transmittance of MEG was due to entrainment of heavies from a Tray Column, supplying methanol to DMC train from another production facility. Heavies reacted with ethylene carbonate in DMC train and product of reaction deteriorated the quality of MEG.

A simulation model was developed for the Tray Column, from where the high boils entrained. Actual performance of the column was evaluated and the results of simulation indicate presence of high boils is not possible under normal circumstances. Methods to prevent the presence of high boils in methanol are discussed in this report, so that production of fiber grade MEG can be sustained.

Estimated production loss was about 7200 MT, equivalent to SR. 5.4 million. Recommendations are given to prevent recurrence of deterioration in MEG quality.

402182 CFD Study of Sieve Tray Dry Pressure Drops

Abstract:

In the chemical processing industry, sieve trays are often used for distillation and absorption columns, but there are few dry pressure drop data available from industrial-sized sieve trays. This paper presents computational fluid dynamics (CFD) and experimental studies of dry pressure drops with commercial-sized sieve trays. The sieve trays studied have the same configurations as the trays used in industrial columns. The effects of hole diameter and deck thickness were investigated using various sieve tray designs. Dry pressure drop predictions from the CFD simulations showed excellent agreement with the dry pressure drop data. CFD simulations showed the velocities were not uniformly distributed across the column cross-section due to the effect of jetting from the deck orifice. Orifice discharge coefficients, derived from the CFD simulations and experimental data, were also compared with those published in literature. In general, the measured and calculated orifice coefficients were higher than the literature model predictions.

Ripple® Trays for Fouling Applications

AIChE 2015 Spring Meeting
Paper # 398200

Steven G. Schon, P.E.
Arkema Inc.

15 AIChE
Spring Meeting, Austin, TX

Introduction

- Tray Fouling
- Approaches to Mitigate Fouling
- Ripple® Trays
- Plant Experience
- Conclusions

15 AIChE
Spring Meeting, Austin, TX

2

Tray Fouling

- **Mechanisms**
  - Thermal Decomposition / Coking
  - Reaction / Polymerization
  - Precipitation
  - Settling
- **Locations**
  - Tray Deck
  - Tray Holes / Valves
  - Underside
  - Downcomers

15 AIChE Spring Meeting, Austin, TX

3

Tray-Deck Fouling

Photo – courtesy of Technip Stone & Webster

15 AIChE Spring Meeting, Austin, TX

4

Tray Orifice (Valve) Fouling

Photo – courtesy of Technip Stone & Webster

15 AIChE Spring Meeting, Austin, TX

5

Under-side Fouling

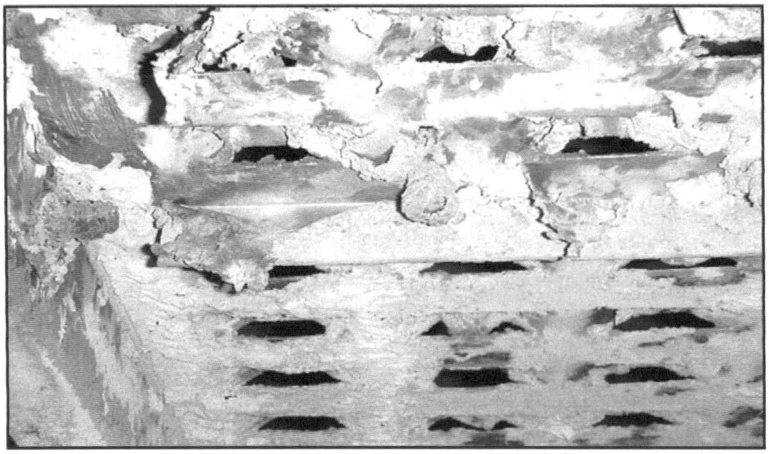

Photo – courtesy of Technip Stone & Webster

15 AIChE Spring Meeting, Austin, TX

6

Fouled Trays With Downcomers

Photo – courtesy of Technip Stone & Webster

15 AIChE Spring Meeting, Austin, TX

7

Approaches to Mitigate Fouling

- **Hydraulics**
  - Increase liquid velocity ("washing" effect)
  - Decrease vapor velocity / increase spacing (entrainment)
- **Additives**
  - Polymerization inhibitors
  - Surfactants
  - Dispersants
- **Materials of Construction**
  - Polished finishes
  - Non-stick coatings (expensive; questionable durability)
- **Fouling-tolerant Geometries**
  - Large orifices / No moving parts (valves)
  - Counter-current flow through holes / eliminate downcomers

15 AIChE Spring Meeting, Austin, TX

8

Ripple® Trays

Photo – courtesy of Technip Stone & Webster

15 AIChE Spring Meeting, Austin, TX

9

Ripple® Trays

- **Counter-Current Contact**
- **No Downcomers**
- **Constant Wetting-Washing Action**
  - Wets underside with downcoming liquid
  - Additives reach underside via liquid
- **Wave Action Promotes Agitation**
  - No stagnant areas on tray deck
  - Reduced Solids Accumulation
- **Polished Finish**
  - Slippery surface to minimize adhesion
    - Punched/drilled holes need to be cleaned to smooth edges

Image – courtesy of Technip Stone & Webster

15 AIChE Spring Meeting, Austin, TX

10

Flow Characteristics
Ripple® vs. Dual-Flow Trays

Dual Flow **Ripple®**

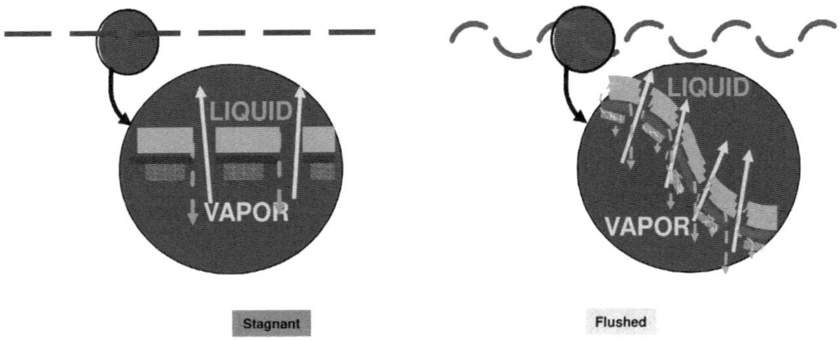

Ripple® Tray Loading Schematic

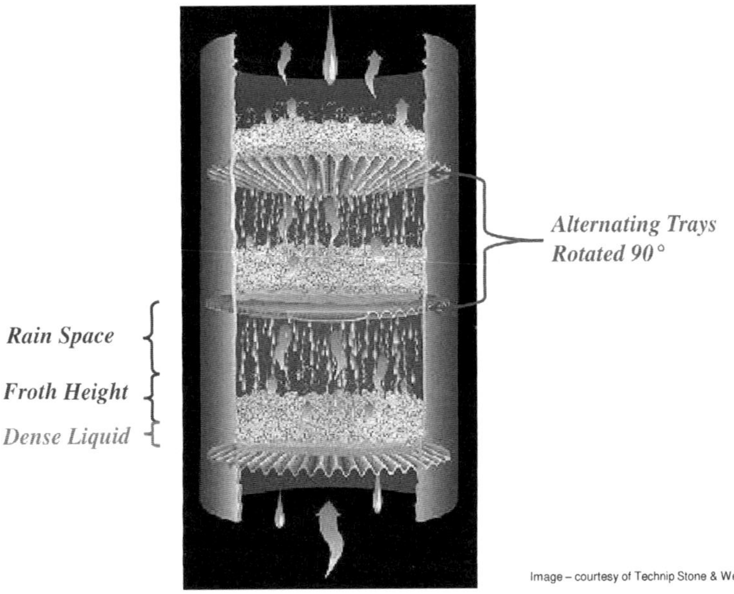

Image – courtesy of Technip Stone & Webster

Plant Experience – Monomer Absorber

- **Hot Monomer-laden Gas**
- **Polymerization Fouling, Especially Under-side**
  - Quenching / absorption of vapor near feed (warm zone)
  - Condensation of un-inhibited hot vapors
- **Original: Sieve Trays**
  - 3-month cleaning cycle
- **Replacement: Ripple® Trays**
  - Bottom section only (sieve trays above)
  - 1½+ years without cleaning

Monomer Absorber After 1 Year With RT

Pan Above Ripple® Tray section

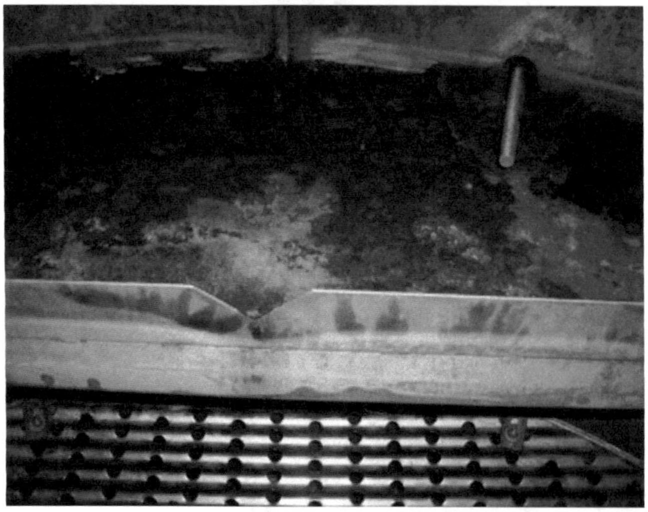

Monomer Absorber After 1 Year With RT

Ripple® Tray – Top Side

15AIChE

15

Fouling in Absorber After 1 Year With RT

Ripple® Tray – Under Side

15AIChE

16

Monomer Absorber After 1 Year With RT

Sump

- Solids accumulated in bottom rather than on trays

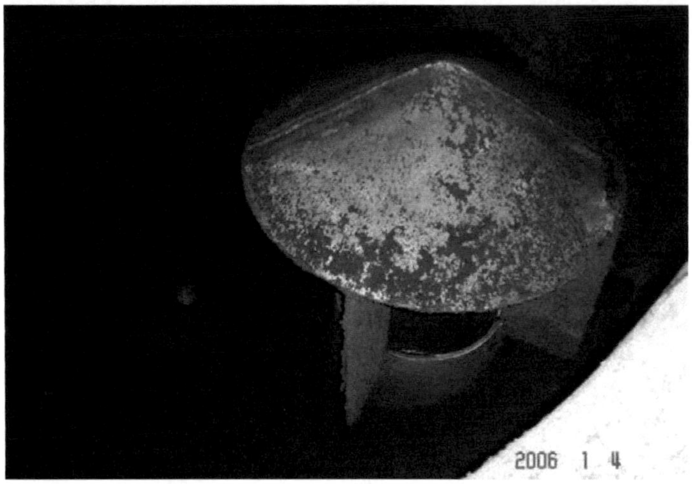

15AIChE Spring Meeting, Austin, TX

17

Other Considerations

- **Tray Efficiency**
  - Low to Moderate for Ripple® Trays
    - **Fractionation**: Better than DualFlow; worse than sieve/valve
    - **Absorption**: Same as DualFlow; better than sieve/valve
- **Capacity**
  - TS&W claims higher capacity than sieve/valve/DualFlow
    - Full active area (no downcomer)
    - Allows closer spacing
- **Turn-down**
  - Limited (~ 2/3 range of sieve & valve trays)
- **Pressure Drop**
  - Higher tray dP than DualFlow or sieve/valve
  - But, if fewer trays (more efficient vs. DualFlow) → lower column dP
- **Cost**
  - Proprietary → relatively expensive
    - Technip Stone & Webster
    - Off-patent; other vendors, but no hydraulic design capability
  - Potential to reduce tower diameter → reduce investment for new unit

15AIChE Spring Meeting, Austin, TX

18

Ripple® Tray Capacity & Turn-down

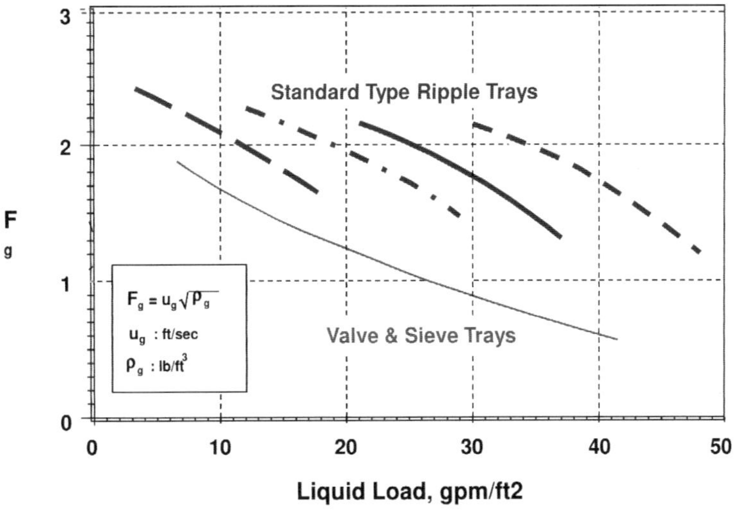

Image – courtesy of Technip Stone & Webster

Ripple® Trays - Conclusions

- **Effective for (Polymerization) Fouling Service**
  - > 6-times longer cleaning interval vs. sieve trays
  - Works best in conjunction with additives / inhibitors
    - Washes underside with additives from downcoming liquid
- **Well-suited for Direct-contact Heat & Mass Txfer**
  - Quench contacting
  - Vapor absorbers
- **Consider for De-bottlenecking in Other Services**
  - Eliminate downcomers (increase active area)
  - Reduce tray spacing / more trays (offset lower efficiency)
  - Vapor-rate limited columns
  - Columns with limited turn-down
- **Relatively Expensive, But Can Have Attractive Payback**
  - **Fouling**: Replace only trays in troublesome section
    - Transition chimney between tray types (redistribute V & L)
  - **Debottleneck**: Full re-tray may avoid need for new larger-ø tower
- **Mature Technology, Though Not Well-known**
  - Over 500 Ripple® Tray towers

NGL Unit Debottlenecked with UOP ECMD™ Trays and High Flux™ Tubing for Significant Production Expansion and Energy Optimization

Shanda G. Fry
Senior Design Engineer
UOP LLC, A Honeywell Company

Joshua M. Moll
Design Engineer
UOP LLC, A Honeywell Company

Patrick Ronan
Process Optimization Engineer
Chevron Phillips Chemical Company LP

Lee Scanlon
Olefins & NGL Technical Services Engineer
Chevron Phillips Chemical Company LP

Prepared for Presentation during the
New Frontiers in Tray Design and Operation Session
2015 AIChE Spring Meeting
Austin, Texas, USA, April 29, 2015

Abstract

UOP high performance distillation and heat transfer equipment have been used extensively in a wide variety of services in the petrochemical and refining industry for more than 40 years. This paper will illustrate the optimization and expansion of Chevron Phillips Chemical's natural gas liquids (NGL) Unit 21 in Sweeny, Texas by coupling the design of new tower internals with new tubing in the associated stab-in reboiler and overhead condenser. Revamping the bottlenecked deethanizer tower with UOP Enhanced Capacity Multiple Downcomer (ECMD™) trays, along with revamping the condenser and stab-in reboiler with UOP High Flux™ tubes, increased unit production rate more than 65% within the existing limitations of the available heating and cooling mediums and plant footprint.

Project Objectives

Capacity at Chevron Phillips Chemical Company LP's Sweeny NGL fractionation complex was increased by approximately 19% in 2013. The project was driven by the increased production of NGLs from shale formations in the region and an opportunity to secure additional long-term NGL supplies into the facility. The NGL fractionation complex consists of multiple units. For economic as well as logistical reasons, the full capacity increase of the expansion project would be achieved at Unit 21, representing a much greater and challenging expansion for that single train. Evaluation of Unit 21 showed that the existing deethanizer would need to be debottlenecked to achieve the targeted capacity increase, while project economics dictated that the existing column and exchanger shells would need to be reused.

In April 2010, Chevron Phillips Chemical contacted UOP to identify an expansion solution for increasing tray capacity and handling the corresponding increase in reboiler and condenser duties. For the expansion project, the following design objectives were given for the modifications to the existing equipment:

- Change operating conditions in tower to maximize capacity (>65% increase) and optimize utility consumption
- Reuse the existing tower with minimal modifications
- Maintain external dimensions of replacement exchangers
- Reuse existing piping and foundation
- Accomplish equipment modifications within the turnaround schedule
- Guarantee performance of the tower system

Post-Revamp Results

The revamp of the deethanizer tower and exchangers with UOP's high performance mass and heat transfer equipment allowed for more than a 65% increase in the feed rate to the column (compared to the maximum pre-expansion feed rate), while maintaining the existing footprint. Table 1 highlights the capacity increase of the expansion project. The data in the table has been normalized using the highest pre-expansion feed rate as the basis (i.e. pre-expansion feed rate is 100%). Additional tower and exchanger capacity (approximately 18%) is expected before flooding occurs.

Table 1. Normalized Operating Data – Before Revamp vs. After Revamp

	Pre-Expansion Operating Data[2]	July 2014 Operating Data[3]
Feed Rate[1]	100	167
Feed Ethane Content, vol%	mid 40s	low 40s
Feed Propane Content, vol%	< 30	> 30
Overhead Product Rate[1]	36	52
Overhead Propane Content, %	2.5	2.5
Bottoms Product Rate[1]	64	115
Observed Reflux Ratio	1.6	2.0
Observed Tray Efficiency, %	-	69
Number of Trays	35	35
Tray Type	1-/2-pass Valve	ECMD Trays
Percent of Flood, %	~100	84

[1] Values have been normalized based on pre-expansion feed rate = 100%.
[2] Normalized values for a 6 hr. period of stable operation on January 13, 2011, representing the highest pre-expansion feed rate while maintaining good cut points.
[3] Normalized values for a 4 hr. period of stable operation from July 14-15, 2014 (post-revamp).

The deethanizer tower was revamped with UOP ECMD trays on a tray-for-tray basis with the column pressure optimized to achieve 90% of flood at the design rates (see Appendix A for more information on UOP ECMD trays). The overhead condenser and bottoms stab-in reboiler were retrofitted with OD coated / ID bare UOP High Flux tubing (see Appendix B for more information on UOP High Flux tubing). This revamp marked the first commercialization of OD coated High Flux tubing with a tube diameter of 15.875 mm.

UOP provided a process guarantee, in addition to a mechanical guarantee, for the new trays and tubes in the deethanizer tower and exchangers. The process guarantee covered the desired feed rate and desired product purities at the design conditions that coincide with operation at 90% of flood.

Modifications to the tower and exchangers were determined to be the most cost effective way to meet the project objectives.

All of the modifications were accomplished within the scheduled project outage of 26 days.

Revamp Design Approach

The key to truly maximizing the achievable production from the deethanizer was to approach the design of the tower and exchangers as an overall unit, and not evaluate each piece of equipment independently. This approach allowed the designers to adjust the process conditions – in this case, operating pressure – to gain the most benefit from all three pieces of equipment until the highest capacity was achieved within the heating and cooling constraints. Many companies have benefited from this approach, which can only be done with the combined use of UOP's ECMD trays and High Flux tubes (McGuire and Sobczyk 2008).

The deethanizer tower originally contained 35 conventional valve trays in a dual-diameter shell. The condenser uses propylene refrigerant to condense the overhead vapor to generate reflux, and the reboiler uses LP / MP steam to partially vaporize the column bottom stream. The pre-expansion deethanizer configuration and equipment details are provided in Figure 1.

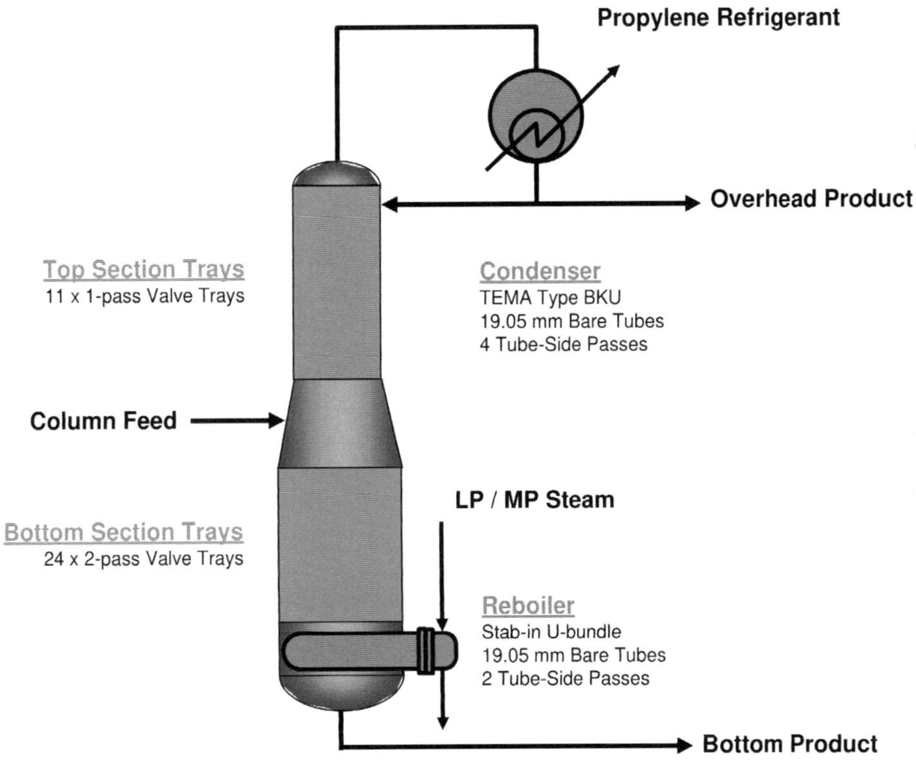

Figure 1. Pre-Expansion Configuration of Deethanizer Unit

Early on in the project, the towers, exchangers, and propylene refrigeration system were identified as bottlenecks for the expansion. To eliminate the need to modify the steam turbine compressor driver, Chevron Phillips Chemical determined they needed to operate the propylene refrigerant 2.32 kg/cm^2 higher than the original design conditions. The increase in the propylene refrigerant pressure would require an increase in the column overhead pressure to obtain a sufficient temperature driving force between the condensing overheads and boiling propylene refrigerant. Unfortunately, the increased operating pressure reduces the capacities of the tower and the reboiler due to smaller delta between liquid and vapor densities, higher reflux ratios required for the given separation at elevated pressures, lower surface tensions, and lower mean temperature difference (MTD) in the reboiler. As a result, when setting the design pressure for the column overhead, there is a trade-off between unloading the condenser system and unloading the trays and reboiler system. To maximize the overall capacity of the deethanizer unit, these competing benefits had to be balanced.

Revamping the deethanizer condenser with High Flux tubes resulted in a minimum MTD for the column overhead vapor and the propylene refrigerant, thus minimizing the column overhead pressure as much as possible. Once the overhead pressure was determined, UOP simulated the column at the desired conditions and confirmed that a tray-for-tray revamp with ECMD trays would not only provide a sufficient number of theoretical stages for the given separation, but would also maximize the column capacity. The original tray count could be maintained considering the tray efficiency of MD and ECMD trays is similar to that of conventional trays in this service and the separation is not very sensitive to changes in tray efficiency (i.e. a 10% loss in tray efficiency only results in a 4% increase in the reflux rate). UOP applied a design efficiency of 65.7%, relative to commercially-available simulation models, which is conservative for ECMD trays in this service (McGuire et al. 1997). Figure 2 shows the efficiency vs. reflux rate curve for the design separation. The x-axis (reflux rate) has been normalized using the design reflux rate as the basis (i.e. design reflux rate is 100%).

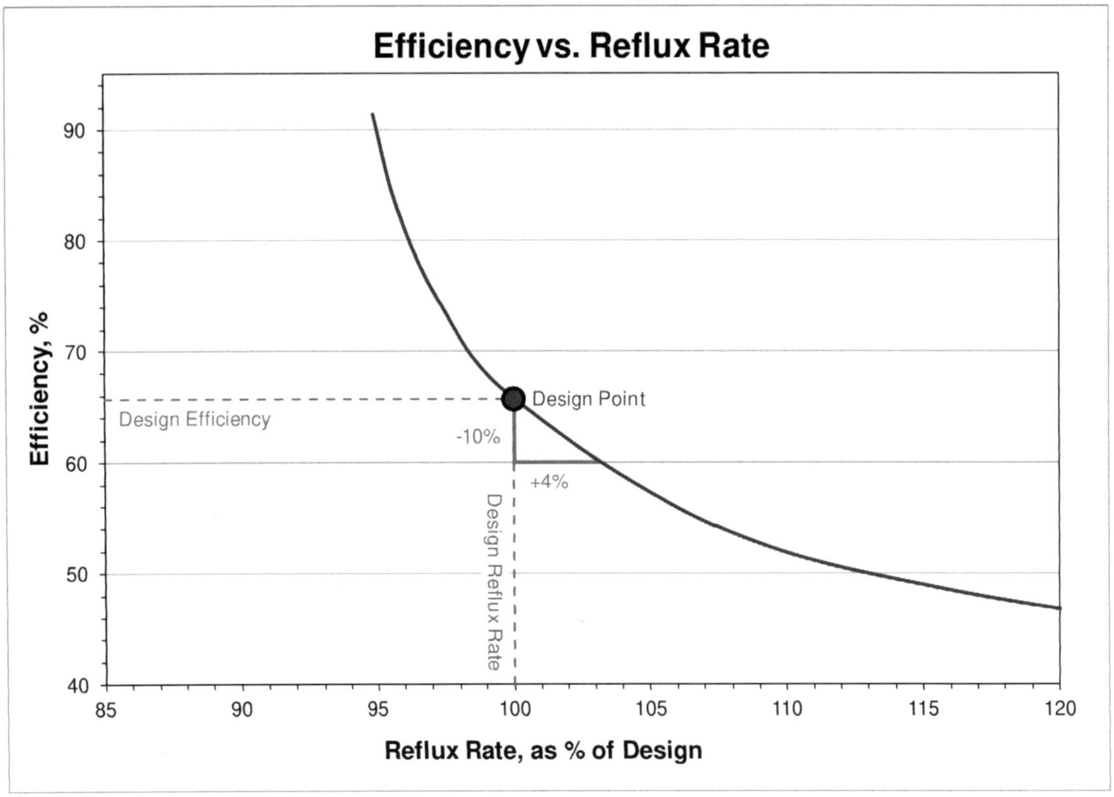

Figure 2. Efficiency vs. Reflux Rate Curve for Design Separation

A sketch of the deethanizer column revamp is shown in Figure 3. A tray-for-tray revamp with 35 ECMD trays allowed for the column capacity to be maximized, relative to a multi-tray revamp configuration.

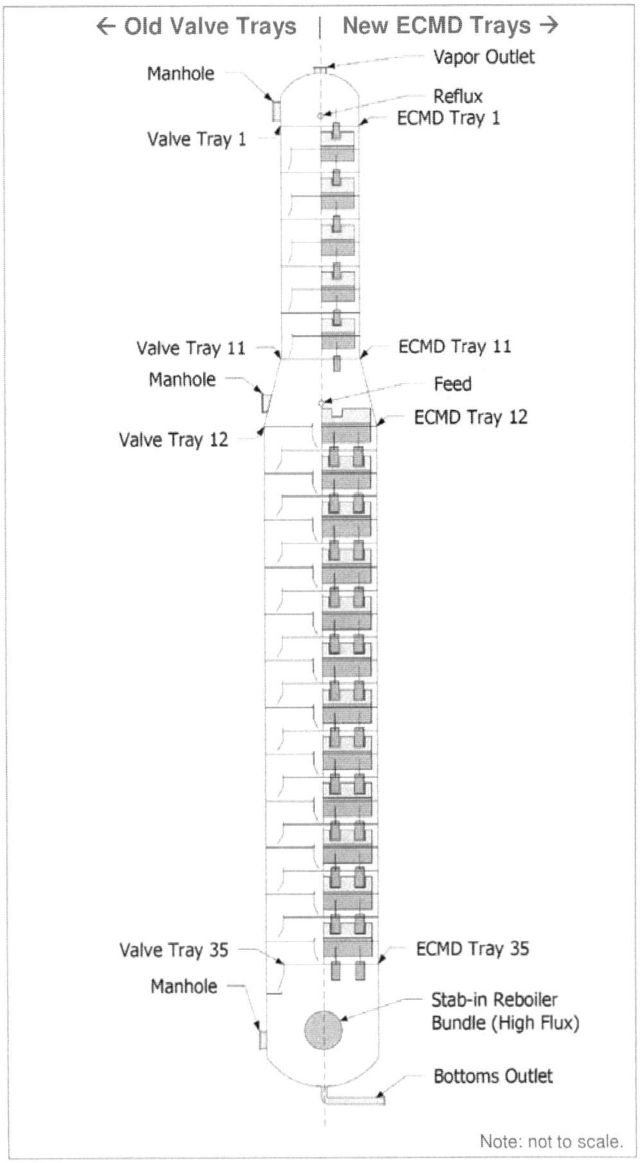

Figure 3. Deethanizer Revamp Sketch

After the tray revamp scheme was determined, UOP evaluated the reboiler and condenser for the new conditions.

The reboiler duty requirement increased by 50% relative to the original bare tube design. The boil-up duty could not be met with the existing bare tube bundle or a 1-for-1 High Flux tube bundle replacement using the existing low pressure steam.

The optimum revamp solution for the deethanizer reboiler involved a maximum area bundle replacement with 15.875 mm diameter OD coated / ID bare High Flux U-tubes oriented on a 20.64 mm square tube pitch. Reducing both the tube diameter and pitch ratio (1.5 to 1.3) allowed a heat transfer area increase of more than 20% without modification to the existing column port diameter. This design allowed the reboiler to stay on low pressure steam, rather than more costly medium pressure steam. Table 2 shows the normalized performance comparison between the original bare tube design and the proposed High Flux tube design.

Table 2. Normalized Reboiler Design Comparison

	Original Bare Tube Design	High Flux Revamp Design
Duty[1]	100%	150%
Area[1]	100%	123%
U-value[1]	100%	146%
MTD, °C	38	32
Percent of Max Bundle Q/A, %	64	97

[1] Values have been normalized based on original 1966 bare tube design = 100%.

The high overall heat transfer coefficient and MTD of this steam driven reboiler warranted a close evaluation of the bundle critical heat flux. In addition, the combination of high reduced pressure, low surface tension, and low latent heat of the boiling fluid contributed to lowering the bundle critical heat flux. Compared with the original bare tube design, the higher duty and lower MTD resulted in a 22% increase in the bundle heat flux. The new calculated bundle heat flux was 97% of the Heat Transfer Research Inc. (HTRI) maximum bundle heat flux and well above HTRI's 70% guideline. This high heat flux, and the substantially larger amount of vapor being generated within the bundle, has the potential to cause liquid starvation and vapor blanketing of the upper tube rows. This condition can reduce both the heat transfer performance and capacity of the exchanger. Therefore, the revamp design incorporated a 110 mm wide vertical bypass lane within the tube layout. This open channel allowed the rapidly boiling vapor to more easily exit the tube field, alleviating bundle critical heat flux concerns. Figure 4 depicts the High Flux bundle containing the vapor by-pass lane.

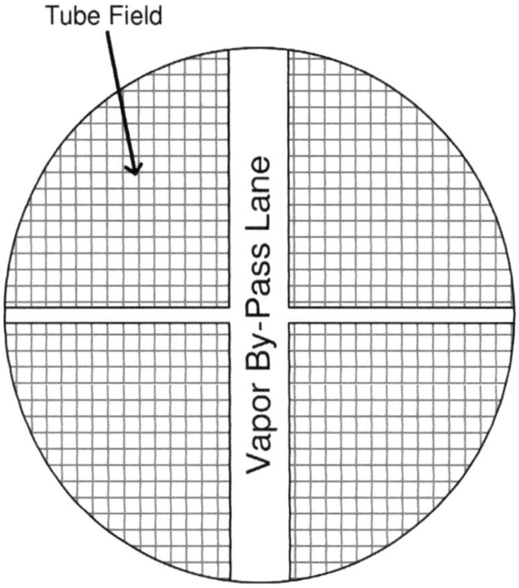

Figure 4. Deethanizer Reboiler Tube Arrangement

The expected condenser duty at 100% of expansion project rates was 70% higher than the original bare tube duty (88% higher at the expected flood point of the ECMD trays), so there was no way the bare tube bundle could accommodate the duty at the target propylene pressure.

UOP determined that a bundle replacement with the same number of 19.05 mm OD coated / ID bare High Flux tubes could accommodate the target duty at the minimum propylene pressure limit (7.24 kg/cm^{2}g).

After the initial exchanger evaluations, there was some iteration regarding allowable product purities (initially tighter specifications were considered to allow for some margin) – they had to be reconsidered based on handling the desired feed rate within the existing column constraints (i.e. column diameter and tray spacing). Once Chevron Phillips Chemical determined the proposed purities would not negatively impact the overall plant production, the system pressure was optimized to prevent the column, condenser, and reboiler from becoming a bottleneck. The proposed tray and reboiler stab-in bundle designs were checked and confirmed acceptable, while improvements were made to the proposed condenser design.

In order to maximize the benefits of the condenser revamp, 15.875 mm diameter tubes were evaluated. Smaller diameter tubes arranged on a tighter tube pitch permitted more heat transfer area to be packed within the existing kettle port diameter. There was some discussion about the allowable tube side pressure drop, since changing to smaller tubes would result in a higher pressure drop for the overhead stream. But since the effect on the reflux temperature was minimal, it was decided that the benefit of increased refrigerant pressure outweighed tube side pressure drop concerns and a pressure drop of 0.7 kg/cm^2 was allowed. The increase in heat transfer area allowed the MTD to be minimized and the propylene boiling pressure to be increased by 0.39 kg/cm^2. The tighter approach temperature was still more than adequate to sustain stable nucleate boiling with UOP's High Flux tubing. The increase in propylene pressure would also further unload the refrigerant compressor by reducing the compression ratio, and lowering the utility costs associated with the propylene refrigeration system.

The optimum condenser revamp involved 15.875 mm OD coated High Flux tubes, arranged on a 19.84 mm triangular pitch. Four tube-side passes and the original tube pitch ratio of 1.25 were maintained. Table 3 details the normalized performance comparison between the original bare tube design and the proposed High Flux tube design.

Table 3. Normalized Condenser Design Comparison

	Original Bare Tube Design	High Flux Revamp Design
Duty[1]	100%	188%[2]
Area[1]	100%	102%
U-value[1]	100%	338%
MTD, °C	7.3	4.0

[1] Values have been normalized based on original 1966 bare tube design = 100%.
[2] The High Flux revamp design duty was 171% of the original bare tube design duty plus 10% additional design margin.

Considering the new duty and rates were nearly 90% higher than the original bare tube design, shell-side liquid entrainment was evaluated for potential impact on the downstream propylene refrigerant compressor. Entrainment calculations showed that maintaining a liquid level at least

70% of the bundle height ensured adequate wetting of the top tube rows with a negligible impact on capacity or performance.

Performance Results

Equipment modifications for the expansion project were completed in early 2013 during the scheduled project outage. The plant successfully started up in the middle of February 2013.

Operating data from July 2014 confirmed that the target feed rate increase had been achieved, with tighter purity specifications than design. Table 4 compares the normalized operating data with the normalized revamp design conditions. Evaluation of the data suggests the efficiency of the ECMD trays is approximately 69%, relative to commercially-available process simulators, which is slightly higher than that of the conventional trays that were replaced. The efficiency vs. reflux rate curve of the operating data is shown in Figure 5. The x-axis (reflux rate) has been normalized using the design reflux rate as the basis (i.e. design reflux rate is 100%).

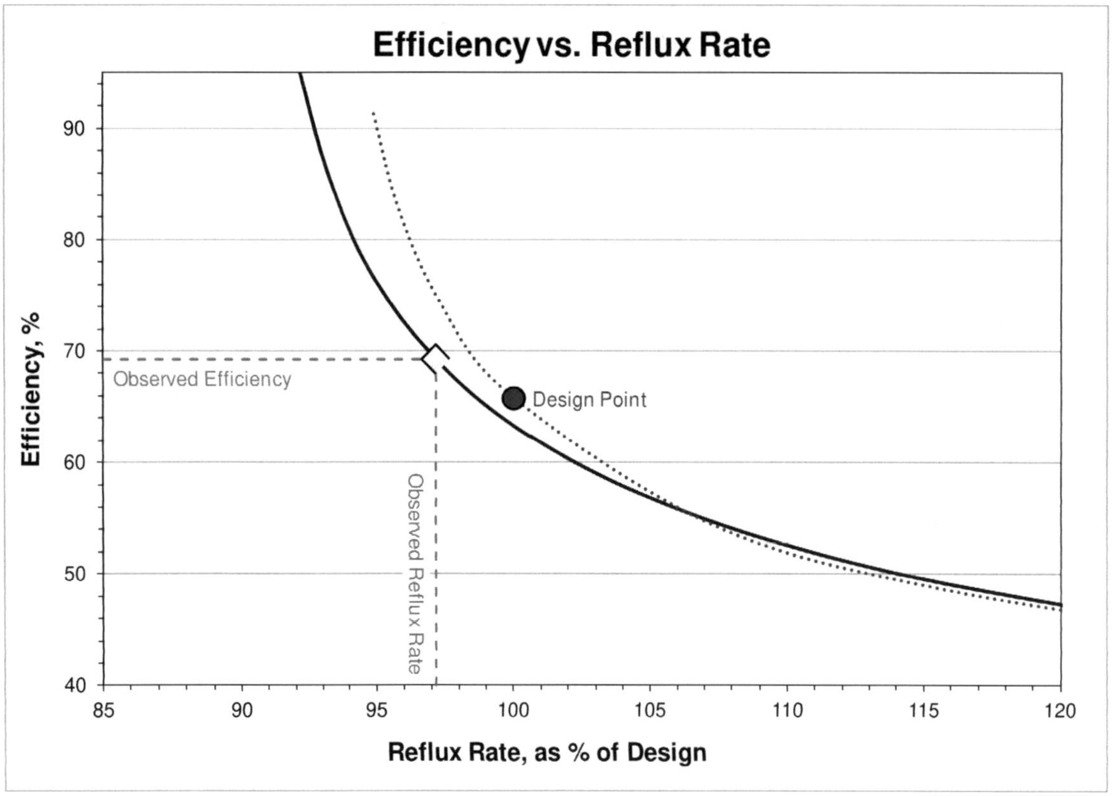

Figure 5. Efficiency vs. Reflux Rate Curves – Design vs. Observed

The higher tray efficiency and changes in the expected feed composition maintained a lower operating pressure. Chevron Phillips Chemical confirmed their ability to push the capacity in excess of 104% of design rates without running at the higher column pressure considered at the design stage. Based on UOP's simulation of the July 2014 operating data, the ECMD trays were operating at about 84% of flood, which meant the trays were expected to process an additional increase of 18% in the feed rate before flooding.

Table 4. Normalized Operating Data vs. Revamp Design

	Pre-Expansion Operating Data[2]	ECMD Tray Revamp Design	July 2014 Operating Data[3]
Feed Rate[1]	100	165	167
Feed Ethane Content, vol%	mid 40s	mid 40s	low 40s
Feed Propane Content, vol%	< 30	< 30	> 30
Overhead Product Rate[1]	36	59	52
Overhead Propane Content, %	2.5	4	2.5
Bottoms Product Rate[1]	64	106	115
Tray Efficiency, %	-	65.7	69.2
Reflux Ratio	1.6	1.8	2.0
Condenser Duty[4]	-	171%	179%
Reboiler Duty[4]	-	150%	146%
Column Pressure Drop[1]	1.0	1.7	1.3
Number of Trays	35	35	35
Tray Type	1-/2-pass Valve	ECMD Trays	ECMD Trays
Percent of Flood, %	~100	90	84

[1] Values have been normalized based on pre-expansion feed rate = 100%.
[2] Normalized values for a 6 hr. period of stable operation on January 13, 2011, representing the highest pre-expansion feed rate while maintaining good cut points.
[3] Normalized values for a 4 hr. period of stable operation from July 14-15, 2014 (post-revamp).
[4] Condenser and reboiler duty ratios based on original 1966 bare tube design duties = 100%.

Figure 6a and Figure 6b show the capacity graphs for the ECMD trays in the top and bottom sections, respectively. The graphs compare the July 2014 operating data with the design point. The x-axis (liquid load) and y-axis (v-load) have been normalized using the design rates as the basis (i.e. design rates are 100%). As previously stated, the lower column operating pressure improved the separation and resulted in increased capacity. The improved separation is a product of a larger delta between liquid and vapor densities, resulting in lower reflux ratios required for the given separation. The increased capacity is a product of increased surface tensions, which allows for higher vapor loads. These phenomena are represented by the lower vapor and liquid loads compared to design conditions and by the different "Vapor Max" and "DC Backup" lines for the design and July 2014 data points.

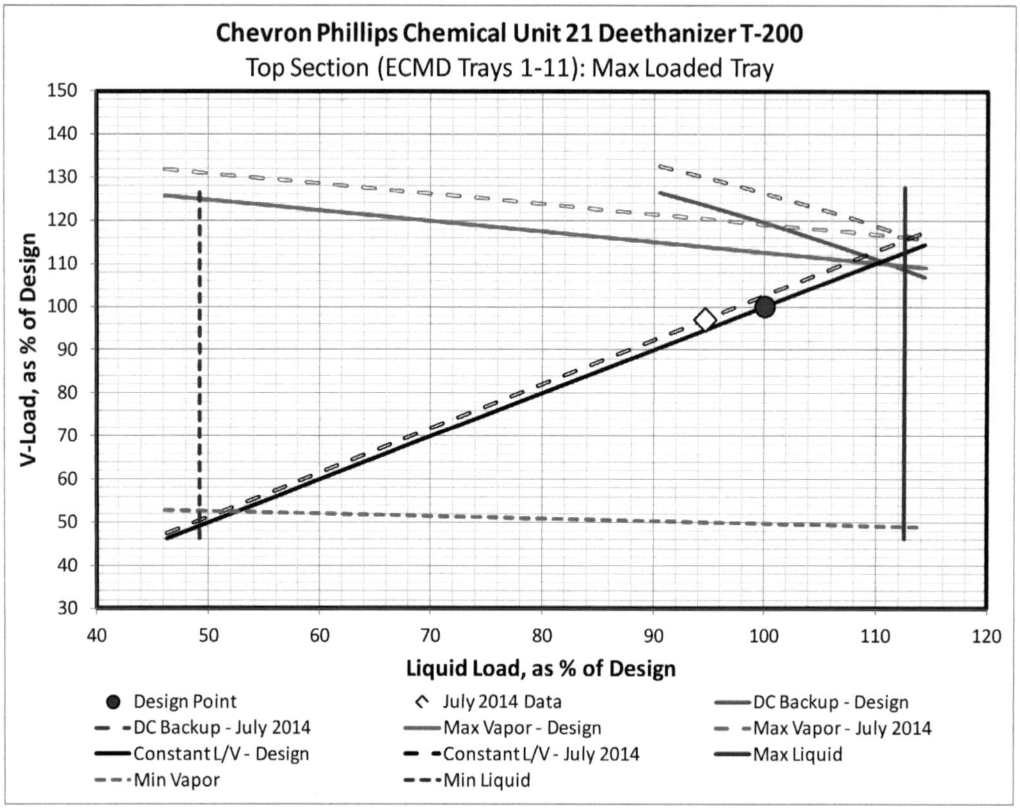

Figure 6a. Capacity Graph – Top Section

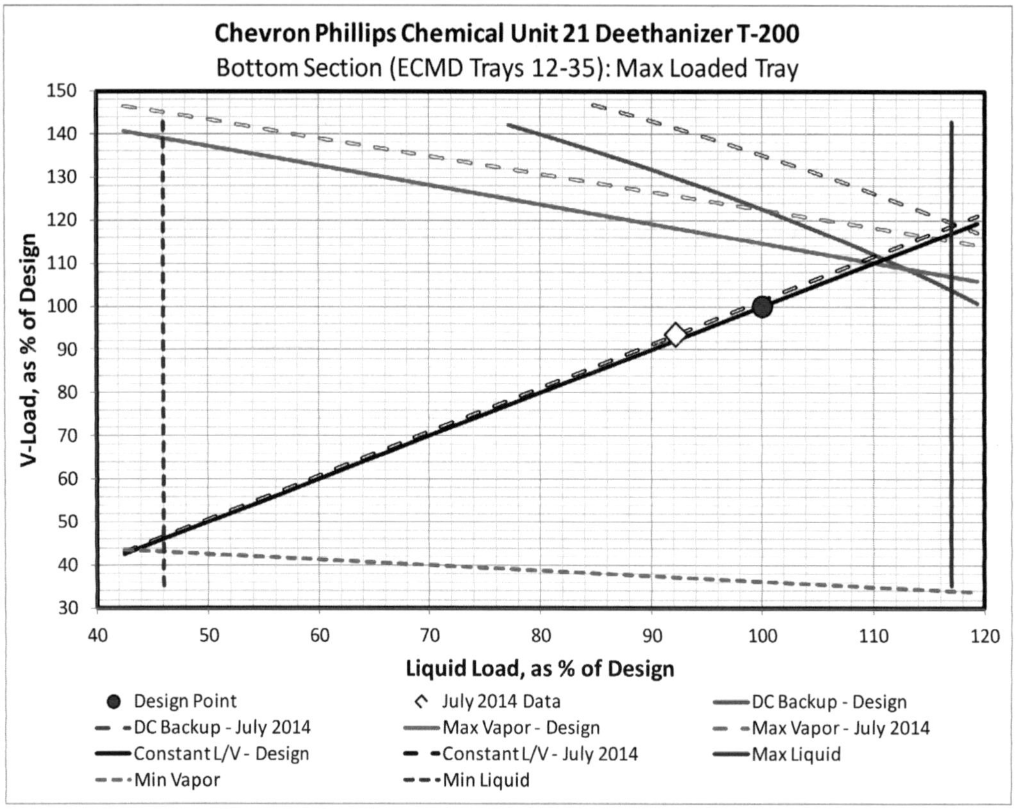

Figure 6b. Capacity Graph – Bottom Section

At one point during the design of the ECMD trays, Chevron Phillips Chemical expressed concerns about the capacity that was offered by UOP regarding the downcomer velocity and jet flood values. In systems that operate at elevated pressures, like the deethanizer, the delta between the liquid and vapor densities is relatively low, making it more difficult for liquid and vapor separation. If not taken into account, this oversight can result in trays that flood earlier than predicted by the different flood correlations.

Based on FRI information, Chevron Phillips Chemical developed maximum criteria for downcomer velocity based on vapor density. For the deethanizer top section with a vapor density of 69.0 kg/m^3, this criteria would have restricted the design downcomer velocity to 220 $m^3/(m^2\text{-hr})$ in order to avoid potential downcomer choking or liquid and vapor separation issues. For the deethanizer bottom section with a vapor density of 87.8 kg/m^3, the system would be limited to an even lower downcomer velocity.

Based on the design rates, the downcomer velocities in the top and bottom sections were close to 375 $m^3/(m^2\text{-hr})$ for the proposed ECMD trays, which greatly exceeded Chevron Phillips Chemical's design criteria for the system conditions, and led them to question the proposed design. UOP stated that their design guideline is around 367 $m^3/(m^2\text{-hr})$ for MD and ECMD trays.

For Chevron Phillips Chemical to be comfortable with the proposed tray design, they required UOP to provide examples of successful operating experience at high downcomer velocities, in a system similar to the deethanizer tower. UOP provided Chevron Phillips Chemical with a plot similar to the one shown in Figure 7 containing the downcomer velocity vs. vapor density for UOP's MD and ECMD tray experience in deethanizer and demethanizer services (i.e. typical services with high vapor densities).

Although the proposed design downcomer velocities were slightly higher than what UOP normally designs for with ECMD trays (considering the vapor density), UOP determined that the higher downcomer velocity was a reasonable trade-off in order to limit the percent of jet flooding to 90%. Based on UOP's in-house data, directionally, the high percent of jet flood constituted more design risk than the high downcomer velocity. Also, UOP had anticipated that there would be a slight reduction in the downcomer velocity when the final design was done, as a result of optimizing the downcomer configuration (i.e. top and bottom downcomer areas).

For the jet flood capacity concerns, Chevron Phillips Chemical was able to confirm that they had past experience with operating ECMD trays at 90% flood and higher at their plants. Consequently, Chevron Phillips Chemical felt that operating the proposed retrofitted deethanizer tower at 90% of flood should be routinely achievable.

During the final design, UOP was able to optimize the tray design to balance the downcomer area and active area in order to lower the downcomer velocities below UOP's design guideline, while keeping the percent of flood less than 90%. The resulting design downcomer velocities in the top and bottom sections were 366 $m^3/(m^2\text{-hr})$ and 367 $m^3/(m^2\text{-hr})$, respectively. These design points are included in Figure 7 for reference.

UOP also plotted the observed downcomer velocities and vapor densities for the ECMD trays based on UOP's simulation of the July 2014 operating data. These values prove that the ECMD trays can handle much higher downcomer velocities than anticipated by conventional design criteria.

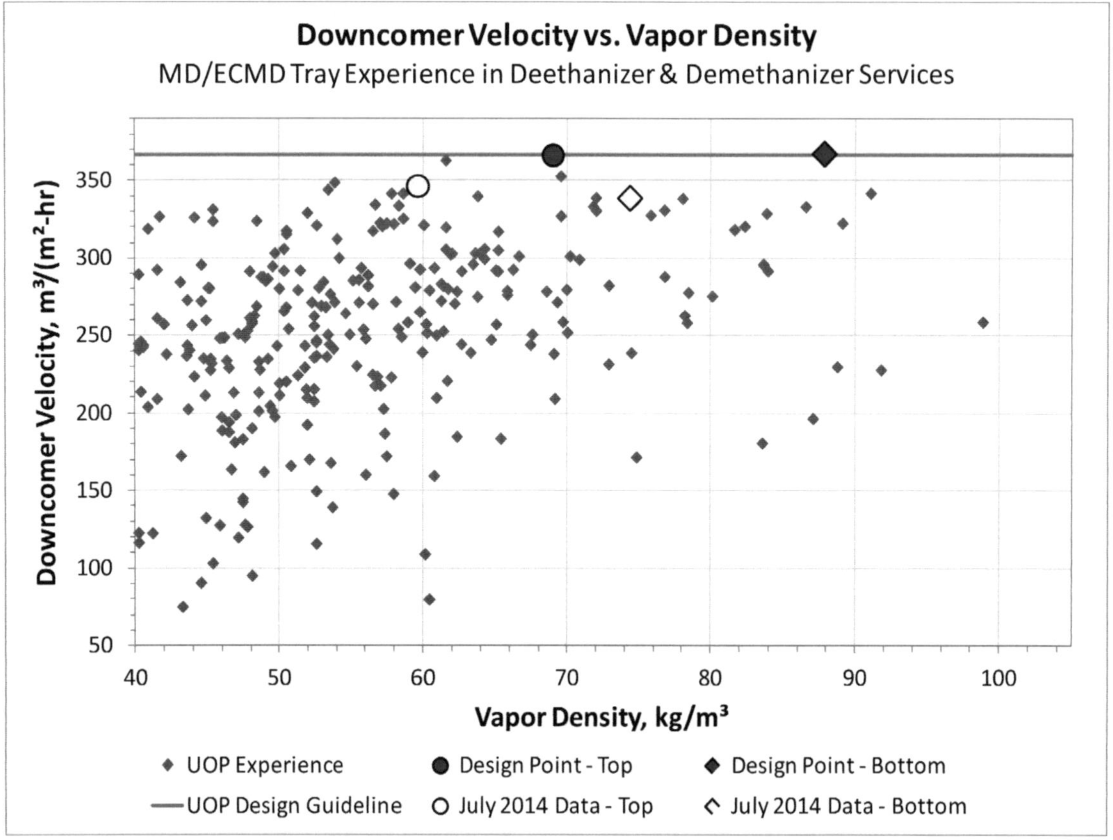

Figure 7. Downcomer Velocity vs. Vapor Density – UOP Experience

The July 2014 operating data also shows that the High Flux bundles are satisfying the expansion requirements. The measured cold-end approach temperature between boiling and condensing streams in the condenser was less than 1°C, and simulations of the operating data indicate the design U-value had been exceeded. Table 5 summarizes the observed condenser performance compared to the High Flux design.

Table 5. Normalized Condenser Performance

	Original Bare Tube Design	High Flux Revamp Design	July 2014 Operating Data[2]
Duty[1]	100%	188%	179%
Area[1]	100%	102%	102%
U-value[1]	100%	338%	394%
MTD, °C	7.3	4.0	3.2

[1] Values have been normalized based on original 1966 bare tube design = 100%.

[2] Normalized values for a 4 hr. period of stable operation from July 14-15, 2014 (post-revamp).

The July 2014 operating data shows that the reboiler is also meeting the goals of the revamp. The LP steam is being throttled to minimize the condensing pressure. The condensate pressure is ~1 kg/cm² less than design. The steam enters superheated and is subcooled to a pinch. Table 6 summarizes the observed reboiler performance compared to the High Flux design. The revamp design was based on the assumption that the steam and condensate would be saturated. The average operating heat transfer coefficient appears lower than design because the steam sensible heat transfer accounts for approximately 15% of the total operating duty.

Table 6. Normalized Reboiler Performance

	Original Bare Tube Design	High Flux Revamp Design	July 2014 Operating Data[2]
Duty[1]	100%	150%	146%
Area[1]	100%	123%	123%
U-value[1]	100%	146%	141%
MTD, °C	38.3	31.9	32.2

[1] Values have been normalized based on original 1966 bare tube design = 100%.
[2] Normalized values for a 4 hr. period of stable operation from July 14-15, 2014 (post-revamp).

Summary

The deethanizer unit at Chevron Phillips Chemical's Sweeny plant was bottlenecked with conventional separation and heat transfer equipment. By revamping with UOP's ECMD trays and High Flux tubing, Chevron Phillips Chemical achieved a capacity increase greater than 65%. At the same time, the implemented changes achieved all of Chevron Phillips Chemical's design objectives. Operating data since the 2013 startup demonstrates that the tower and exchangers are operating better than design.

As a result of the revamp, both Chevron Phillips Chemical and UOP are pleased with the performance of the Unit 21 deethanizer system. The revamp of the deethanizer unit adds to the long history of working with Chevron Phillips Chemical to maximize the benefits from operating their various distillation units located in different plants throughout the country. Chevron Phillips Chemical has also used UOP heat and mass transfer equipment in other areas of the Sweeny complex (Shakur et al. 1999).

References

1. P. J. McGuire and M. R. Sobczyk, "Maximizing Return on Assets and Process Optimization of New Units with UOP Equipment Technologies" 2008.

2. P. J. McGuire, M. S. M. Shakur, A. G. Cantrell, J. Valverde, and D. Webb, "Replacing MD™ Trays in Amoco Deethanizer Yields 20% Capacity Increase" American Institute of Chemical Engineers Spring National Meeting, Houston, Texas, March 12, 1997.

3. M. S. M. Shakur, R. E. Tucker, K. J. Richardson, M. R. Sobczyk, R. D. Prickett, C. Polito, and S. E. Harper, "Increase C2 Splitter Capacity with ECMD Trays and High Flux Tubing" American Institute of Chemical Engineers Spring National Meeting, Houston, Texas, March 18, 1999.

Appendix A – MD and ECMD Tray Product Description

For more than forty years, UOP has been a leader in supplying high performance distillation technology, including MD and ECMD trays. These trays have been used in over 1000 applications, and the performance and benefits of these trays are discussed below.

MD and ECMD trays are designed for medium to high pressure service, where the liquid to vapor ratio is relatively high. Whereas conventional trays must go to two-pass, four-pass, and even six-pass trays to accommodate these high liquid rates, the unique design of the MD and ECMD trays make them ideally suited for these applications.

An MD tray consists of a series of long, trough like downcomers, separated by sections of perforated plate (see Figure A-1). These downcomers are terminated in the vapor space of the tray below, and are spouted to allow liquid to flow directly onto the panel of the tray below eliminating the need for receiving pans. This configuration allows for larger active area and downcomer area, thus increasing the tray capacity.

Figure A-1. MD Tray Configuration

The most unique feature of the MD tray is the large number of downcomers, resulting in a very long outlet weir length. This extended weir length reduces the liquid height on the tray, and as vapor rises through this thinner liquid layer, lower froth heights are generated compared to conventional trays. This lowers the percent of jet flood and further increases the capacity. The lower froth height also allows these trays to be designed for a lower dry tray pressure drop, as lower vapor velocities are required to prevent the liquid from weeping through the decks.

The combination of no receiving pans, lower froth heights, and lower dry tray pressure drops all serve to significantly reduce the pressure drop of the MD tray. In fact, the pressure drop of an MD tray is only about 60 percent of that of a comparable conventional tray. As the pressure drop across the tray makes up the largest component of the downcomer backup, MD trays have much lower downcomer backups than conventional trays. The combination of low froth heights and low downcomer back-ups allows the trays to be installed at much smaller tray spacings than conventional trays. In the case of revamps, installing a larger number of trays at smaller tray spacings can significantly increase the number of theoretical trays generated within the column, allowing for improved product purities or reduced energy requirements for a given separation. In new units, installing the trays at smaller tray spacings can result in a significant reduction in overall column height.

Figure A-2 shows a side view of an MD tray. The box-like structure of the downcomers provides all of the mechanical support, and no additional bolting bars or beams are required. This photo also shows another unique feature of the MD tray, that consecutive trays are rotated ninety degrees. As shown in Figure A-3, the crossing of the downcomers on consecutive trays forms a "unit cell", and all of these cells on the tray perform exactly the same as the others. As the diameter of the column increases, the number of downcomers also increases, such that the unit cell dimensions remain relatively constant. This allows the MD tray to be scaled up to very large diameters, with no fall-off in efficiency or capacity, which is often experienced with conventional trays at large diameters.

Figure A-2. 90 Degree Tray Rotation

Figure A-3. Unit Cell Geometry

The crossing of the downcomers also results in the MD tray being self-balancing, i.e., the trays themselves are able to correct a maldistribution problem. If there is a problem in the column that results in all of the liquid being distributed to one side of the column, the downcomers will act very quickly to redistribute the liquid evenly across the tray, resulting in maximum capacity and efficiency.

For 30 years, the MD tray was the industry standard for high capacity trays. However, in the late 1980's, a need developed for an even higher capacity tray, in order to revamp columns that already contained MD trays, as well as for future expandability for customers who wanted to use MD trays in new units. To address this need, UOP developed the ECMD tray.

Figure A-4. ECMD Tray Configuration

An ECMD tray is very similar to an MD tray in configuration, but differs in two main aspects (see Figure A-4). An ECMD tray uses slotted decks to introduce some of the vapor horizontally into the liquid to reduce the froth height further, providing for increased vapor capacity. Disengagement plates are used over the downcomers to redirect the slot-propelled froth into the downcomer, and to enhance vapor disengagement. These trays provide a 15-20% capacity increase over MD trays, while maintaining the same separation efficiency, and have been installed in hundreds of columns since 1990.

Other UOP Mass Transfer Products

UOP offers a full range of mass transfer products for various high capacity and high efficiency applications (see below). This paper is focused on the benefits provided by one of our high capacity tray products. The other products also provide many benefits to customers looking to increase / improve production of existing columns, or looking for capital savings on the installation of grassroots columns.

- Family of MD Trays
  - **Multiple Downcomer (MD) Trays** – High Capacity

    [+30-50% more capacity than conventional trays]
  - **Enhanced Capacity (ECMD) Trays** – Higher Capacity

    [+45-70% more capacity than conventional trays]
  - **Parallel Flow (PFMD) Trays** – High Efficiency & High Capacity

    [+45-70% more capacity than conventional trays with similar / better efficiency]
- **SimulFlow Devices** – Highest Capacity

  [+105%-130% more capacity than conventional trays]
- **Slotted Sieve Trays** – High Efficiency & Low Pressure Drop
- **Multiple Upcomer (MU) Trays** – Liquid-Liquid Extraction

ECMD Trays

MD Trays

PFMD Trays

Slotted Sieve Trays

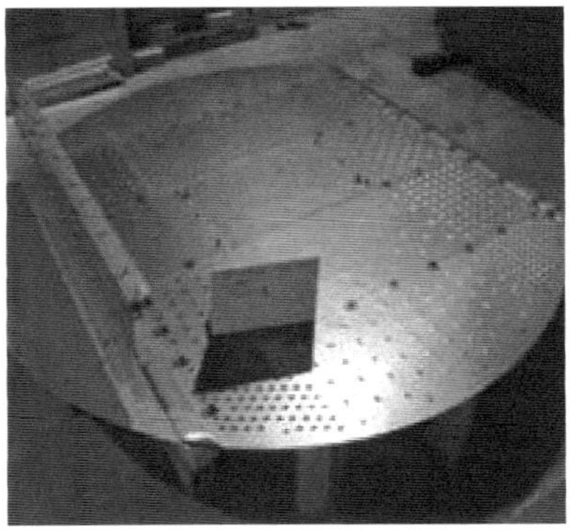

SimulFlow Devices

MU Trays

Appendix B – High Flux Tube Product Description

High Flux tubing features a nucleate boiling enhancement that significantly increases the heat transfer performance of shell and tube heat exchangers. The surface consists of a porous metal coating metallurgically bonded to the inside or outside surface of a bare tube. With conventional tubing, boiling begins randomly and usually at pits and scratches in the surface. With High Flux tubing, capillary action draws liquid into the porous layer and vaporizes from the large number of nucleation sites that are integral to the structure. These sites allow vapor bubbles to be generated at a wall superheat of less than 1°C. This is considerably lower than that required to initiate nucleate boiling on the surface of a plain tube. Vigorous nucleate boiling, at low wall superheat, results from a combination of good matrix thermal conductivity, highly extended microsurface area, and a large number of interconnecting channels within the porous surface. As a result, the boiling heat transfer coefficient improves by a factor of ten or more. This produces overall heat transfer coefficients that are up to two to five times higher than conventional tubing in boiling applications. The relative performance improvement will be a function of the shell side and tube side fluids, exchanger geometry, and design fouling factors for the specific application.

See Figure B-1 for a visual boiling comparison for High Flux, bare, and low fin tubes.

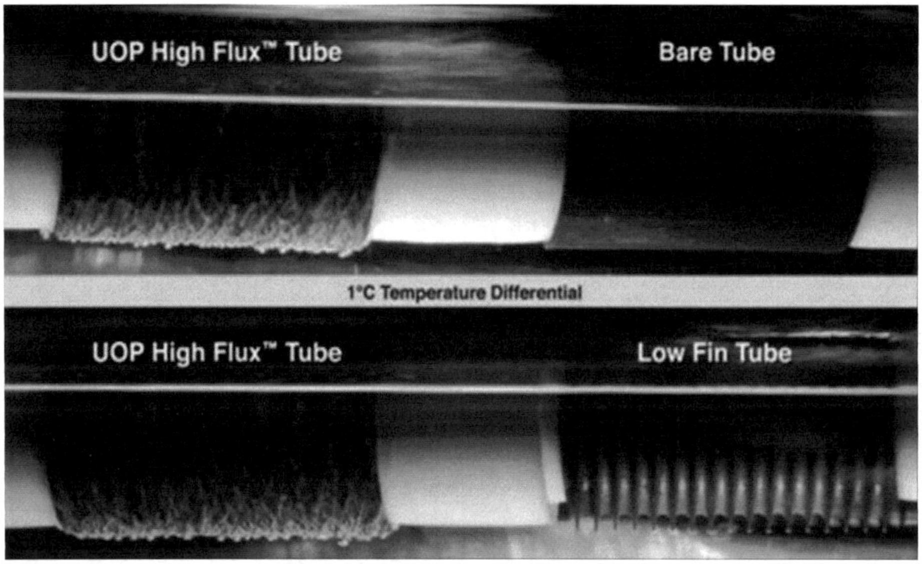

Figure B-1. Pool Boiling Test

High Flux tubing is available in most common materials such as carbon steel, stainless steel, copper alloys and aluminum, with the coating being the same material as the tubular substrate. The manufacturing process ensures a mechanically strong surface highly resistant to abrasion or erosion. The porous coating on one tube surface can also be combined with sensible- and condensing-side enhancements (finning or fluting) on the other surface to produce a doubly-enhanced High Flux tube to meet performance requirements. Horizontal thermosyphons, kettles and stab-in reboilers, with boiling taking place on the shell side of the exchanger, use OD coated tubing with lengths manufactured up to 15240 mm. Vertical thermosyphon reboilers use ID coated tubes with lengths up to 6100 mm.

UOP's experience in refinery and petrochemical applications dates back more than 40 years. Ideal High Flux exchanger candidates typically use condensing hydrocarbons or steam to vaporize steam or hydrocarbon streams. Reduction in capital investment, reduction in the number of exchanger shells, and efficient operation at small temperature differences are some reasons cited by operating companies or contractors for selection of High Flux heat exchangers.

Figure B-2 shows a photograph of the High Flux tube with surface magnification.

Figure B-2. High Flux Boiling Surface

Other UOP Heat Transfer Products

UOP offers a full range of enhanced heat transfer products for boiling and condensing applications (see photo below). This paper was focused on the benefits provided by one of our tube products. Similar benefits can be offered by the other products, as needed to satisfy a specific application.

UOP Heat Transfer Tube Products

High Flux and High Cond Tubes

399982 Channel Baffles on Trays

Abstract:

Frequently distillation tray designers are posed with the problem of handling moderate to high volumes of liquid with small amounts of vapor. Side Strippers, Sour Water Strippers and Wastewater Strippers are typical examples of such applications where this problem frequently arises. Tray performance is critical in these services and a shortfall in performance could have severe detrimental effects downstream. Conventional wisdom is to provide blanking on the tray or "dead zones" where no contacting occurs. A better practice is to provide Channel Baffles which contract the active area of the tray without reducing the flow path length of the device. The impact of channel baffles on tray performance was measured with a water cooling technique in Aston University's 2.44 m diameter air-water simulator. Temperature isotherms (analogous to concentration profiles) of the developed flow patterns on a channel baffled sieve tray are compared to the same circular sieve tray. This paper will discuss the issues shown above and describe good Channel Baffle design practices that one can employ to ensure that a good, low vapor loaded, design is achieved.

402185 Reverse Vapor Cross Flow Channeling in High-Capacity Trays: Diagnosis and Fixup

Abstract:

This paper originates from a systematic troubleshooting effort that successfully diagnosed two complex independent problems which severely limited the capacity of an atmospheric crude distillation tower. The crude tower had been retrayed (by others) with modern high-capacity trays, but upon start-up achieved less than 80% of its previous capacity.

The troubleshooting identified two bottlenecks: premature flooding on some of the heavy naphtha-diesel fractionation trays, and a phenomenon which we termed "excursion," in which sudden temperature reductions around the diesel draw, accompanied by distillate yield loss, occurred upon rate increase.

Our previous paper (Kister, H. Z., N. O'Shea, and D. Cronin, "Systematic Field Tests Turn Loss into Gain in High-Capacity Trays, Part 1: Excursion", Proceedings of the Distillation Topical Conference of the AIChE National Spring Meeting, Houston, TX, April 2007) focused on diagnosing and fixup of the excursion bottleneck. This paper focuses on identifying and fixup of the high-capacity trays bottleneck.

Multichordal gamma scans with quantitative analysis (Kister, H. Z., "Use Quantitative Gamma Scans to Troubleshoot Maldistribution on Trays", Chem. Eng. Progr., February 2013) provided detailed mapping of the froth heights, froth densities, hydraulic heads, and entrainment profiles on the trays. These profiles showed conclusively that the trays were bottlenecked by a previously unknown mechanism which we termed "Reverse Vapor cross Flow Channeling (RVCFC)". This phenomenon is closely related to vapor cross flow channeling (VCFC), but is induced by the forward push on high capacity trays rather than by the hydraulic gradient on the trays. Like VCFC, this phenomenon sets in when there is a significant channeling initiator while the friction pressure drop of the trays is too low to counter the channeling and when the ratio of flowpath length to tray spacing is high.

Based on this diagnosis, the tower was retrayed with the same type of tray but with a design that gave significantly larger friction pressure drop. The new trays also had two passes which halved the ratio of flowpath length to tray spacing. The modifications completely eliminated the bottleneck and the RVCFC, permitting higher-than-design capacities to be achieved.

402183 Should You Consider Moving to Fixed Valves?

Abstract:

In recent years we have seen a general shift from floating valves towards using fixed valves on distillation trays. This is particularly the case in the refining industry where the loss of some operational flexibility is often considered an acceptable tradeoff for the benefits that fixed valves bring, including better reliability, fouling resistance and more robust designs that can better withstand operational upsets. This paper will present a review of the different types of fixed devices and their benefits including R&D and CFD development work and industrial case studies.

402184 Design of Tray Columns for Emergency Blow-Down Situations

Abstract:

The process and hydraulic design of column internals is done at a very early stage of a project. At this time neither blow-down rates nor a pressure safety valve (PSV)-design is available. Therefore, blow-down cases are typically not considered in the tray design. This may lead to the situation that the forces on column internals, due to increased flow rate and pressure drop, are exceeding the mechanical design limits during blow-down.

To eliminate the risk of mechanical damage to the column internals during blow-down situations – which leads to repair costs and additional downtime – a detailed investigation of the blow-down situation is required. The flow rates passing through the column internals during blow-down can be multiples of the normal operating flow rates. Thus, the balance between an acceptable pressure drop at operating conditions, avoiding damage to the trays at blowdown, and an economic design is challenging.

Therefore, for specific applications, Linde Engineering performs fluid dynamic calculations for the blow-down situation of the plant or plant section. As a result, transient flow rates and physical properties during blow-down can be obtained. These are used to perform calculations for every column section to identify its criticality. To obtain transient pressure profiles, additional tray by tray calculations may be performed. Considering the tray and mass-transfer device geometry, the resulting forces on the equipment are calculated. Based on this, FEM calculations of the trays using the 3D CAD/FEM software SOLIDWORKS are performed. Results are the mechanical stress and resulting deformation of the tray segments, bolts and clamps.

The load on clamps and tray segments may cause damage to the parts directly or can cause that tray segments are torn out of their fixation. Every column internal being identified as mechanically overloaded will be further investigated. The design of the trays is adapted to cover blow-down situations without mechanical damage.

A crucial point in the often time-critical situations of a project (execution) is the combination of mechanical and hydraulic design tools and FEM software, which allows finding an optimal design within the available time frame.

The described workflow was successfully followed in the design of trays for a gas plant. A reinforced tray design for the blow-down situation was developed for several columns. The workflow and the obtained results are further discussed in the presented paper.

401084 Optimal Design and Synthesis of Ternary Distillation Systems Involving Complex Column Configurations

Abstract:

The synthesis of the optimal separation sequence for the distillation of multicomponent mixtures is until today a challenging task [1] particularly if complex column configurations are involved. Indeed, the available short-cut methods used for the screening of different complex distillation column alternatives including single-feed columns with a side stream above or below the feed section, columns with side stripper / rectifier, and side stream columns with prefractionator are approximate and have many restrictions [2-4].

In this contribution, the evaluation of the above mentioned complex distillation designs for the separation of ternary mixtures is carried out at operating conditions with respect to both the total vapor flow rate provided by the column reboiler(s) [3-4], and the minimum total number of theoretical stages of the respective column (s), so that both relevant screening criteria [3] related to operating and capital costs, may be considered.

For this purpose, a new generalized design approach [5] for the determination of the minimum internal vapor / liquid flow rates at pinch condition for each column configuration, is applied. The new design approach assumes constant relative volatilities and constant molar flow rates within each column section. Furthermore, it is based on the analytical characterization of each mass transfer section as well as on the identification of all possible pairs of constant composition zones in a column including both those around the feed sections and those around the side-stream sections so that no conceptual splitting [3, 6] of a column with side stripper / rectifier in a sequence of simple columns is required. Accordingly, Underwood's equations may be applied to each pair of constant composition zones along with the corresponding separation taking place in the two respective column sections should be considered. For the calculation of the minimum total number of theoretical stages of a column at operating condition, optimization algorithms based on the analytical formulations of the distillation line, the Eigenfunction and the number of theoretical stages for each column section, have been elaborated with taking into account the mass balance at each feed section [7]. Furthermore, the analysis of sharp separation geometry for a column with a side stripper / rectifier allowed the determination of the optimal composition of the column side stream (feed of the side section) at operating conditions. Indeed, it was found that for a column with a side stripper, the optimal composition of the liquid side stream (feed of the side stripper) is located within the straight line section bound between the saddle pinch of the rectifying (or middle) section and the unstable node pinch of the side stripping section. Likewise, for a column with a side rectifier, the optimal liquid composition corresponding to the vapor side stream (feed of the side rectifier) under prevailing thermodynamic equilibrium, is located within the straight line section bound between the saddle pinch of the stripping (or middle) section and the stable node pinch of the side rectifying section.

As compared to the available short-cut methods, the new design methods require a minimum number of specifications, apply for arbitrary feed quality, are easy to use, and do not need any graphical support [8].

The design methods developed in this work for the investigated complex column configurations, have been applied for the calculation of design parameters for separation examples including ternary ideal behaving mixtures, as well as ternary azeotropic mixtures approximated by appropriate pseudo-ideal mixtures [7]. For all tested separation examples, the developed design algorithms have been successfully validated for each complex column configuration by comparing the relevant design parameters obtained in this work with those obtained by rigorous simulation using Radfrac (Aspen plus), so that a very good agreement between them could be established.

Finally, the new short-cut methods have been successfully, used for rapid and accurate screening of the above presented complex column configurations as well as the direct / indirect simple column sequences; allowing thereby the synthesis of the optimal complex column configuration / sequence for the distillation of both ideal behaving and azeotropic ternary mixtures.

References

(1) Shenvi AA, Shah VH, and Agrawal R. New Multicomponent Distillation Configurations with Simultaneous Heat and Mass Integration, AIChE Journal. 2013, 59, 1, 272-282.

(2) Tedder DW and Rudd DF. Parametric studies in industrial distillation: part I. Design comparisons, AIChE Journal 1978, 24, 2, 303-315.

(3) Glinos K and Malone MF. Optimality Regions for Complex Column Alternatives in Distillation Systems, Chem. Eng. Res. Des. 1988, 66, 229-240.

(4) Nikolaides IP and Malone M.F. Approximate Design of Multiple-Feed/Side-Stream Distillation Systems. Ind. Eng. Chem. Res. 1987, 26, 1839-1845.

(5) Adiche C. and B. Ait Aissa, A Generalized Approach for the Conceptual Design of Distillation Columns with Complex Configurations, (submitted, 2014).

(6) Carlberg NA and Westerberg AW. Temperature-Heat Diagrams for Complex Columns.

2. Underwood's Method for Side Strippers and Enrichers. Ind. Eng. Chem. Res., 1989, 28, 1379-1386.

(7) Adiche C and Vogelpohl A. Short-cut Methods for the Optimal Design of Simple and Complex Distillation columns. Chem. Eng. Res. Des. Part A. 2011, 89, 1321-1332.

(8) Rooks RE, Malone MF, Doherty MF. A Geometric Design Method for Side-Stream Distillation Columns. Ind. Eng. Chem. Res. 1996, 35, 3653-3664.

397863 Crude Unit Heat Integration - Feed Composition Effects and Current Best Practices

Abstract:

Heat integration provides the bulk of the energy required for atmospheric distillation of crude oil. Heat integration effectiveness must balance multiple factors. These factors include;

1. Energy costs.
2. Capital costs.
3. Operability and control.
4. Integration with other units and shutdown scheduling.
5. Corrosion.
6. Feed flexibility.
7. Product types and conditions.

The emphasis of this paper is on the analysis of a well heat-integrated atmospheric-vacuum distillation unit. Comparisons for different types of crude oils covering the range from 20 to 50 API degrees are shown. These comparisons identify best-practice targets for heat integration in crude units running a variety of feeds.

399409 Dividing Wall Distillation Steady State Model Validation

Abstract:

Dividing Wall Distillation Steady State Model Validation

Bailee Roach, Bruce Eldridge

James R. Fair Process Science and Technology Center

University of Texas at Austin;

Patented in 1949, the dividing wall distillation column (DWC) is slowly gaining traction in the United States. In 2007, it was estimated there were more than 100 DWC's in industrial use, an increase from 60 in 2002. The benefits of a DWC versus traditional separation methods are staggering, including estimated energy savings of 30 percent. Despite these advantages, predictive models are required to develop an understanding of column operation. This insight will promote the widespread adoption of DWC technology.

The University of Texas, in collaboration with industrial partners, has constructed a pilot scale DWC. Outfitted with structured packing, the six inch diameter pilot column was designed to allow for a wide range of operability and process control studies. A steady state study has been performed separating a tertiary mixture of hydrocarbons. Data from the experimental pilot column was used to validate a steady state simulation. Both MPC and PID control were employed and the optimal control strategies were determined.

402190 Fluidized Packed Beds (FPB) – an Alternative for Process Gas Treatment in "Dirty" Systems?

Abstract:

Columns with fixed internals, such as random or structured packings, are well known and widely used in fluid separation processing. However, there are applications where the immobilized packings fail or cause serious operational problems due to their susceptibility to clogging during the treatment of solid loaded media.

Fluidized packed beds are discussed as an alternative for gas-liquid operations in systems with a high solid content. Continuous movement of the packings allows self-cleaning of the packing and thus promises stable operation even in systems with very high solid content. Unless the possible advantage in "dirty" systems, fluidized packed beds are far less common compared to fixed packed beds, this is partly due to the lack of reliable design data for this type of apparatus.

In order to design a fluidized bed packed column, experimental investigations of the fluid dynamics and mass transfer behavior have to be performed in order to generate design principles. In this work, the results of extensive experimental investigation of fluidized beds are carried out in a specially designed test plant with a diameter of ds = 0.6 m. The hydraulics of fluidized beds made of polypropylene have been investigated using the air/water system under ambient conditions in the gas load range up to the flooding point and at specific liquid loads between 0 and 80 m3/(m2h). Particular attention was paid to the determination of the fluidization point, characterizing the lower limit of the operating range of the fluidized bed. The start of fluidization of the packing is crucial to characterize the operating range of the fluidized bed. Based on the SBD model [1], a model equation has been developed on the basis of the experiments, allowing a prediction of the fluidization point and pressure drop of the wetted bed. The separation efficiency and the effective mass transfer area of the fluidized bed were determined based on experiments with different standard test systems, including the system CO2/air-NaOH under ambient conditions. The results were discussed in detail in the context of literature data and compared with conventional Pall rings.

Additionally, specific challenges in design and scale-up of fluidized beds are addressed.

The design data developed in this work has been used for the design of an industrial plant for H2S absorption from flue gas, experiences in design, fabrication and operation of the industrial plant are presented as well.

From the presented work it can be concluded, that fluidized packed beds may be a good alternative for gas-liquid operations like scrubbing and stripping in systems with a high solid content.

402189 Investigation of Fluid Dynamic Parameters in Packed Distillation Columns Using X-Ray Computed Tomography

Abstract:

The flow behavior of two liquid phases in packed distillation columns is hardly understood. The final aim of this work is to develop a model for the design and process optimization of three-phase distillation, using reliable data for the flow behavior of two liquid phases on a packing.

In this work the X-ray computed tomography (CT) is used as non-invasive measuring technique. The experiments are made with a customized computer tomograph at our institute that enables measurements of flows in vertical objects. The detector and the X ray source are rotating around the column and the column is not moved, so that the flow inside is not influenced by centrifugal force. As a result of the high spatial (70 μm) and temporal (1000 projections per second) resolution, dynamic processes inside separation columns can be monitored with a high precision. Fluid dynamic parameters in distillation columns (like interfacial area, liquid hold-up and liquid film thickness on the packing or between two liquid phases) can be directly estimated from the recorded CT images of a packing cross section at different height positions along the column. For the evaluation several steps of image treatment were applied to finally obtain segmented images with specific gray values for each component.

The experiments were performed in a glass column with an inner diameter of 100 mm and a packing height of 1000 mm consisting of five packing segments. For the investigation of the fluid dynamics five different special designs of Mellapak 500.Y packing from Sulzer Chemtech AG were used. The special designs are a smooth, fluted and gridded texture with holes and a smooth and gridded texture without holes. Additionally the influence of the rotation of the packing segments to each other by 0° and 90° on the accumulation of the liquid between two packing segments and the liquid distribution along the packing height was determined. For the investigation of the influence of the substance properties on the flow morphology and the fluid dynamics, different liquids with low and high viscosities were used. The liquids were fed with a baffle distributor and the liquid load was varied between 5 and 50 m^2 / m^3 h. In a former study we showed that the packing is not evenly wetted and the liquid film thickness is not constant over the entire packing. Hence we developed a method to determine the liquid distribution and the flow morphology on a packing cross section and the packing height.

For the study on the flow behavior of the two immiscible liquids, water and n heptane were fed into the column with a baffle type distributor with two separate feeding points for each liquid to ensure that the two phases are evenly distributed on the packing. The flow behavior of the two liquids on the packing was analyzed by varying the total liquid load between 5 and 70 m^3 / m^2 h. The feed ratio of water and n-heptane was 1:1, 2:1 and 3:1. Furthermore the flow morphology of the interacting liquids was investigated using a three dimensional visualization of short packing sections along the column.

ANALOGY OF MASS TRANSFER IN ABSORPTION AND DISTILLATION COLUMNS: WETTED WALL COLUMN STUDY

Jan Haidl[a], Frantisek J. Rejl[a], Lukas Valenz[a], Vaclav Linek[a] and Michael Schultes[b]
[a] *University of Chemistry and Technology, Prague, Czech Republic*
[b]*RASCHIG GmbH, Ludwigshafen am Rhein, Germany*

Introduction

The absorption and distillation performed in packed columns can be treated as analogical processes. In both processes the liquid flows down on the packing where it is in contact with the gas/vapor and the mass transfer occurs between the phases. The volumetric mass-transfer coefficients, k_La and k_Ga, are usually used as the mass-transfer rate parameters which relate the component flux through the gas/liquid interface and the mass-transfer driving force – difference in component concentration at the interface and in the phase bulk. The values of the mass-transfer coefficients depend on the system properties and process conditions. If the assumption of the process analogy is valid it should be possible to build identical relations for the mass-transfer coefficients calculation which would be applicable for both processes.

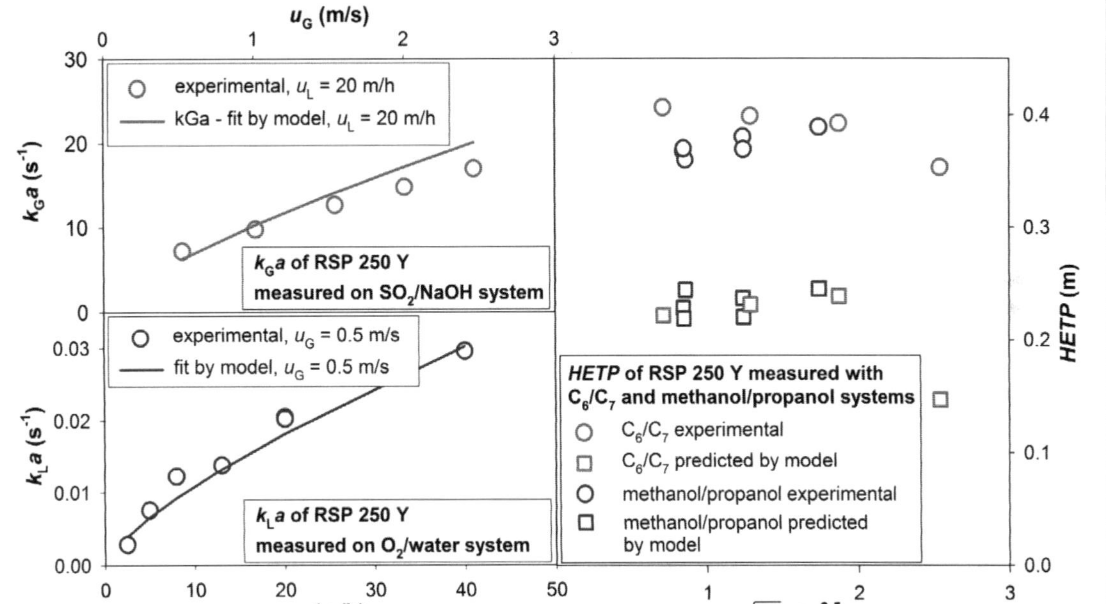

Fig. 1. Comparison of measured *HETP* values of RSP 250-Y packing with values calculated with model of Billet and Schultes[1] adjusted to the results of absorption experiments

According to the principle of the universal model the mass-transfer characteristics measured under absorption conditions can be re-calculated to the distillation ones providing reliable estimation of the packing separation performance (usually in form of *HETP*). However as it is shown in Fig. 1, when the Billet and Schultes[1] model* is adjusted to fit well the absorption mass-transfer coefficients, then

* The model of Billet and Schultes (1999) is chosen as it is derived as an universal mass-transfer model for absorption and distillation and enables to fit its parameters to the experimental data

it seriously underestimates the *HETP* values for distillation systems. This discrepancy lays the questions: is it really possible to calculate the distillation performance out of the absorption data and vice versa and if so then how to do it.
This study is focused on the possibility of transfer the mass-transfer coefficients between absorption and distillation. It will be performed on the device which provides identical interfacial area for both processes eliminating its possible influence on the mass-transfer coefficients.

Furthermore, being aware of the fact that the physical properties of standard absorption systems usually differ considerably from properties of standard distillation systems we chose to use such absorption systems and conditions which are similar to the distillation ones. For the scope of this paper the process conditions and system properties are mostly characterized by dimensionless Reynolds and Schmidt numbers of the phases. As the mixtures of primary alcohols are considered as systems suitable for distillation experiments, those alcohols were also used for the k_L measurements. The properties of the gas phase for the k_G measurements were varied using three gases differing considerably in density.

In this paper the mass-transfer characteristics of the wetted wall column measured under absorption conditions utilizing systems with properties close to those of distillation systems are presented. These characteristics will be in further work used for the prediction of the column distillation performance in terms of *HETP* and mass-transfer coefficients and compared to the experimentally determined values.

Experimental setup and procedures

The experiments were performed in two glass columns of 25 mm inner diameter (geometrical area a_G = 160 m^2/m^3) and lengths of 1.4 m (k_L measurements) and 2.1 m (k_G and a_e measurements). The columns are equipped with 5 (7) pairs of holes which enable placing the liquid and gas sampling devices along the column. The liquid is fed to the column by means of the needle distributor which is capable to create a stable film for the flow rates higher than 110 ml/min for pure water and 80 ml/min for NaOH solution and organic solvents. Schematic drawing of the setup is shown in Figs. 2 and 3.

k_L measurements

The k_L was measured by desorption of O_2 from the liquid into the stream of nitrogen. The absorption liquid (water, methanol, ethanol, n-propanol, n-heptane and equimolar mixture of methanol and *n*-propanol) was saturated with pure oxygen at desired temperature in the tempered and well stirred vessel and fed to the column. An additional glass exchanger was installed right in front of the liquid distributor to ensure the liquid entering the column has the desired temperature. Flow rates from 60 up to 600 ml/min and temperatures from 20 to 75 °C were utilized. The liquid samples were analyzed by means of the optical probe (distance between the sampling points, h = 0.6 m). The k_L was calculated according to the Eq. 1.

$$k_L = \frac{Q_L}{\pi \cdot d \cdot h} \ln\left(\frac{p_{O_2,top}}{p_{O_2,bottom}}\right) \qquad (1)$$

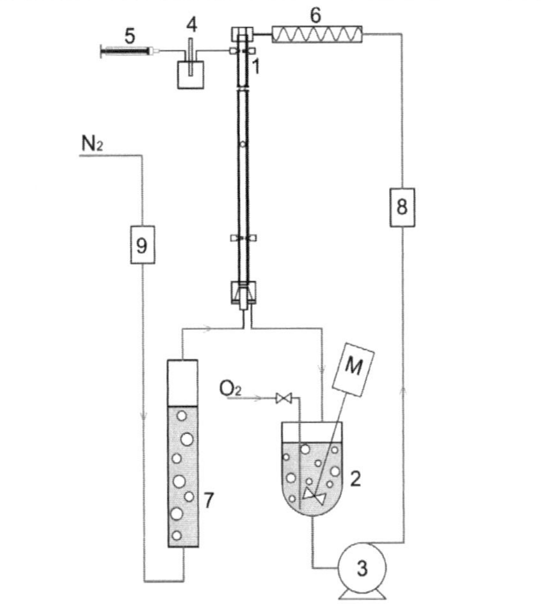

1 - glass column, 2 - stirred vessel, 3 - gear pump, 4 - measuring cell, 5 - syringe pump, 6 - glass heat exchanger, 7 - bubbled column, 8 - liquid flow meter, 9 - gas flow meter

Fig. 2. Experimental setup for the absorption k_L measurements

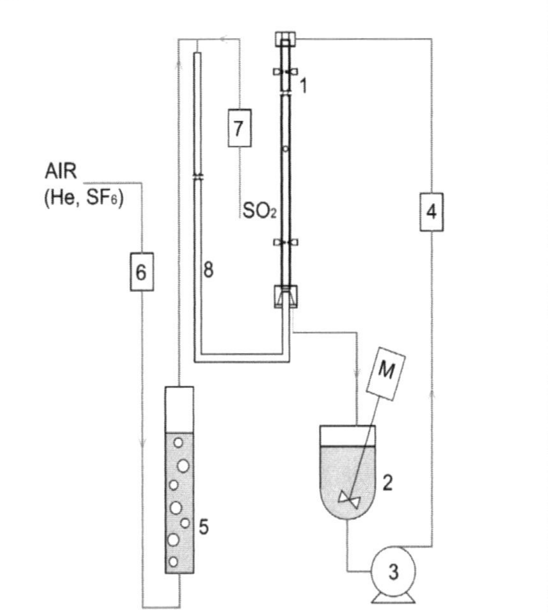

1 - glass column, 2 - stirred vessel, 3 - gear pump, 4 - liquid flow meter, 5 - bubbled column, 6 - gas flow meter, 7 - tracer gas flow meter, 8 - mixing tube

Fig. 3. Experimental setup for the absorption k_G and a_e measurements

k_G measurements

The k_G measurements were measured using the standard SO_2/NaOH system. Three carrier gases (air, He, SF_6) saturated with the vapors of water were utilized providing wide range of the gas phase physical properties (Sc_G = 0.45 – 2.75). SO_2 was injected into the carrier gas in a 2.5 m long mixing tube before entering the column. The gas samples collected from the column (sample points distance, h = 1.8 m) were analyzed by IR spectrometer. Eq. 2 was used for the k_G calculation.

$$k_G = \frac{u_G \cdot d}{4 \cdot h} \ln\left(\frac{c_{SO_2,bottom}}{c_{SO_2,top}}\right) \quad (2)$$

a_e measurements

The effective interfacial area, a_e, was measured using the air/CO_2/NaOH system. The same column setup as for the k_G measurement was used, the gas samples were analyzed by the IR spectrometer. The concentration of the NaOH (1 kmol/m^3) was determined by titration with HCl solution. The effective interfacial area was calculated by the Eq. 3. The parameter of K_a = 0.00243 m/s related to 1 kmol/m^3 NaOH solution was determined independently by still level experiments.

$$a_e = \frac{u_G}{h \cdot K_a} \ln\left(\frac{c_{CO_2,bottom}}{c_{CO_2,top}}\right) \quad (3)$$

The conditions utilized for the measurements were chosen with respect to the conditions expected in the distillation column: Re_L = 100 – 400, $Sc_L \approx 130$, Re_G = 5000 – 20000, Sc_G = 0.5 – 1.

Results

Effective Interfacial area

Effective interfacial area of the column, a_{eff}, measured with CO_2/NaOH system has been found 158 -161 m^2/m^3 for several gas and liquid velocities which is in excellent agreement with the geometrical area of the column (a_G = 160 m^2/m^3).

Mass transfer in the liquid phase

The k_L data acquired for all organic solvents at all temperatures for $Re_L > 80$ have been correlated in dimensionless form by Eq. 4

$$Sh_L = 0.33 Re_L^{-0.12} Sc_L^{0.47} \qquad (4)$$

with RSD of 6 %. The Eq. 4 is valid for Re_L = 80 – 600 and Sc_L = 66 – 1200 and covers the range of conditions expected in distillation experiments.

As there is no reliable source for the diffusivities of oxygen in organic solvents at various temperatures, the diffusivities have been estimated by Tyn-Calus method as recommended by Reid et al.[2]

Our Sh_L data has been compared with those of Hikita et al.[3] and Won and Mills[4] which were measured with non-aqueous systems. The data of Hikita were found to fit well to our correlation, whereas the data of Won and Mills were found to be approx. 20 % lower then predicted from our correlation.

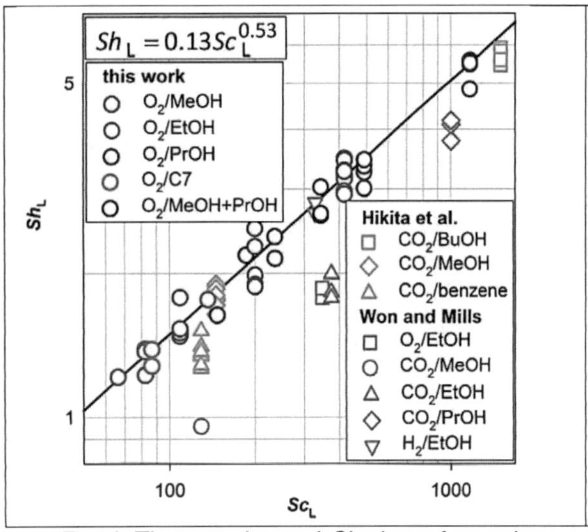

Fig. 4. The experimental Sh_L data of organic systems as a function of Sc_L. Comparison to data of Hikita et al.[3] and Won and Mills[4] obtained with non-aqueous systems

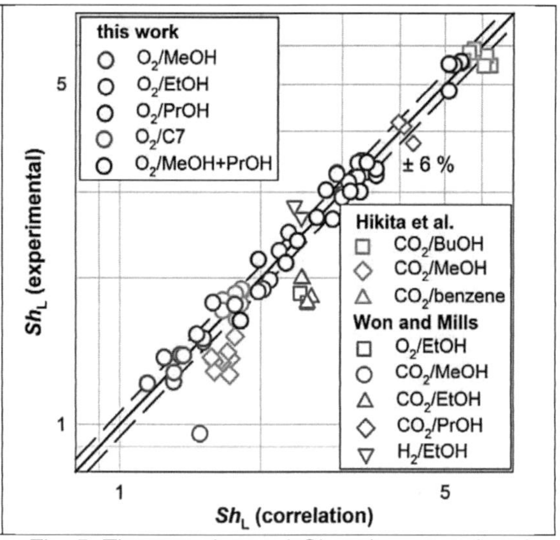

Fig. 5. The experimental Sh_L values correlated by Eq. 4 as a function of Re_L and Sc_L. Comparison to data of Hikita et al.[3] and Won and Mills[4] obtained with non-aqueous systems

The shape of the correlation determines $k_L \propto D^{0.53}$ close to the prediction of the Higgbie's penetration theory. The mass-transfer data in terms of Sh_L obtained for different solvents, flows and temperatures are fitted well by the correlation and mutually overlap, supporting the rationality of their description in the terms of chosen dimensionless numbers Re_L and Sc_L. General success of the correlation suggests that in the range of Reynolds and Schmidt numbers used it shall be valid for similar organic solvents. Agreement with the Hikita's data measured with carbon dioxide suggests its applicability also for different solutes.

Mass transfer in the gas phase

The mass transfer in the gas phase of the wetted-wall column has been studied by chemisorption of SO_2 from the air, He, and SF_6 into the aqueous solution of sodium hydroxide. The k_G data measured on all three systems were correlated together in dimensionless form resulting in Eq. 5 which is valid for Re_{LG} = 1200 - 12000 and Sc_G in the range 0.45-2.75 with RSD of 4% (see Fig. 7).

$$Sh_G = 0.012 Re_{LG}^{0.90} Sc_G^{0.61} \qquad (5)$$

In the range of its validity the correlation agrees well with the results of Gilliland and Sherwood[5] and within 20 % also with those obtained by Crause and Nieuwoudt[6] and Johnstone and Pigford[7] (see Fig. 7).

The power of Reynolds criteria 0.9 falls in between 0.68-1.05 found by the other investigators. The correlation provides 0.61 power of the Schmidt number, which is higher than that found by Gilliland and Sherwood[5] and also higher than 1/3 utilized in Chilton-Colburn analogy. The value is close, however, to the 0.566 value found by Crause and Nieuwoudt[6].

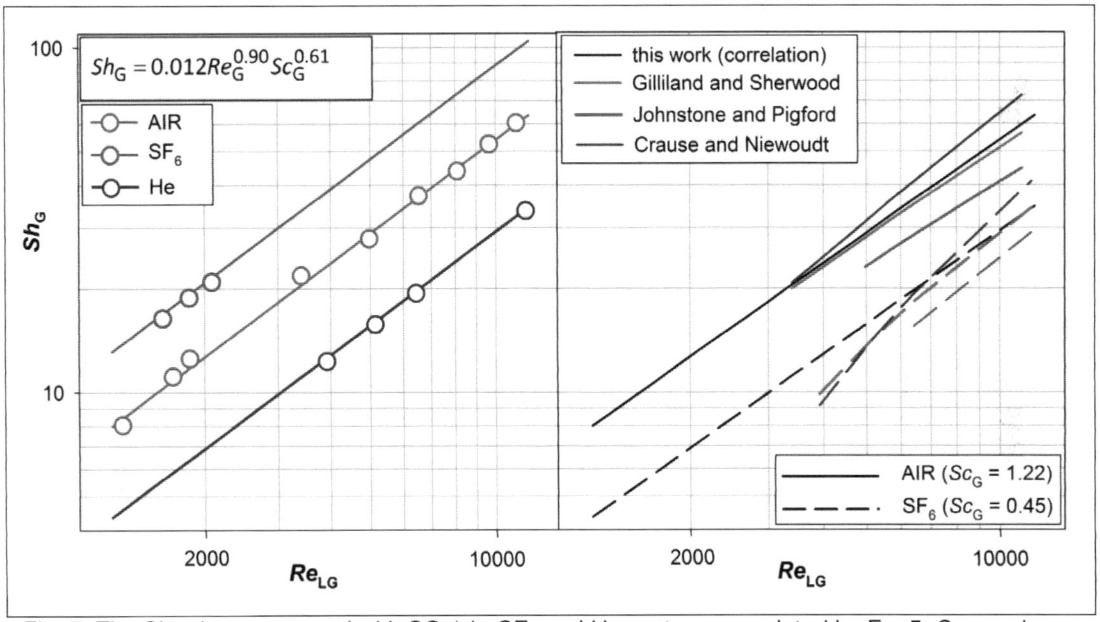

Fig. 7. The Sh_G data measured with SO_2/air, SF_6 and He systems correlated by Eq. 5. Comparison with correlations of Gilliland and Sherwood[5], Johnstone and Pigford[7] and Crause and Nieuwoud[6]

Conclusions

This paper brings mass-transfer data of the organic liquid phase and of the several carrier gasses in a wetted wall column for absorption/desorption conditions. Such data have been only rarely provided in the literature.

There was shown that obtained correlations of the mass-transfer data in terms of the dimensionless *Sh*, *Re* and *Sc* numbers

$$Sh_L = 0.33 Re_L^{-0.12} Sc_L^{0.47}$$

$$Sh_G = 0.012 Re_{LG}^{0.90} Sc_G^{0.61}$$

fit the mass-transfer data for any of the studied system and therefore seem to be universal. Both correlations seem to agree well with available literature data, the highest difference between experimental and literature data (30 %) was found for the k_L data measured by Won and Mills[4] with O_2/ethanol and CO_2/ethanol systems.

The aim of the work has been to obtain absorption data under conditions approaching those found for atmospheric distillation of alcoholic systems in the same column. Comparison of the data will be used in the broader scope of study of the transfer of the data between absorption and distillation. The revealed correlations predict relative resistance to the mass-transfer in the liquid up to 45 % (methanol-*n*-propanol system) which shall be sufficient for its more precise determination.

Acknowledgement

Authors are grateful to Raschig GmbH for financial support of the project, to the Czech Science Foundation for supporting the project through Grant 13-01251S, and to the financial support from specific university research (MSMT No 20/2015).

References

1. Billet, R. and Schultes, M. (1999), „Prediction of Mass Transfer Columns with Dumped and Arranged Packings. Updates Summary of the Calculation Method of Billet and Schultes.", in *Trans IChemE*, 77 (Part A), p. 498-504.

2. Reid, R.C., Prausnitz, J.M., Poling, B.E. (1987), „The Properties of Gases and Liquids, 4.th ed.", *MacGraw Hill, New York.*

3. Hikita, H., Nakaniski, K., Kataoka, T. (1959), „Liquid-phase mass transfer in wetted-wall columns.", in *Kagaku Kogaku*, 23, p. 459-466.

4. Won, Y. S. and Mills, A. F. (1982), „Correlation of the effects of viscosity and surface tension on gas absorption rates into freely falling turbulent liquid films.", in *International Journal of Heat and Mass Transfer*, 25(2), p. 223-229.

5. Gilliland, E.R. and Sherwood, T.K. (1934), „Diffusion of Vapors into Air Streams", in *Industrial and engineering chemistry*, 26(5), p. 516-523.

6. Crause, J.C. and Nieuwoudt, I. (1999), „Mass Transfer in a Short Wetted-Wall Column. 1. Pure Components", in *Ind. Eng. Chem. Res.*, 38, p. 4928-4932.

7. Johnstone, H.F. and Pigford, R.L. (1942), „Distillation in a wetted-wall column", in *Transactions of American Institute of Chemical Engineers*, 38, p. 25-51.

Symbols

a_e	effective interfacial area of the column, $m^2 \cdot m^{-3}$
a_G	geometrical area of the column, $a_G = 4/d$, $m^2 \cdot m^{-3}$
c	molar concentration, $mol \cdot m^{-3}$
D	diffusivity, $m^2 \cdot s^{-1}$
d	column diameter, m
g	gravity acceleration, 9.82 $m \cdot s^{-2}$
h	height of the column, distance between sampling points, m
K_a	reactive-diffusion parameter in Eq. 3.3, $K_a = 0.00243$ m/s

k_G	gas phase mass-transfer coefficient, m/s
$k_G a$	volumetric gas phase mass-transfer coefficient, s^{-1}
k_L	liquid phase mass-transfer coefficient, m/s
$k_L a$	volumetric liquid phase mass-transfer coefficient, s^{-1}
p	pressure, Pa
Q	volumetric flow rate, $m^3 \cdot s^{-1}$
u_G	gas velocity, m/s
u_L	liquid film velocity according to the Nusselt's theory, $u_L = \left(\frac{Re_L^2 \mu_L g}{48 \rho_L} \right)^{1/3}$, m/s
u_{LG}	gas veloctity relative to the liquid suface, $u_{LG} = u_L + u_G$, m/s

Greek symbols

μ	viscosity, Pa·s
ρ	density, $kg \cdot m^{-3}$

Subscripts

bottom	bottom position
G	gas phase
L	liquid phase
top	top position

Dimensionless numbers

$Re_L = \frac{4 \cdot Q_L \cdot \rho_L}{\pi \cdot d \cdot \mu_L}$	liquid phase Reynolds number
$Re_{LG} = \frac{d \cdot u_{LG} \cdot \rho_G}{\mu_G}$	gas phase Reynolds number relative to the liquid surface
$Sc = \frac{\mu}{\rho \cdot D_{12}}$	Schmidt number
$Sh_G = \frac{k_G d}{D_{12,G}}$	gas phase Sherwood number
$Sh_L = \frac{k_L}{D_{12,L}} \left(\frac{\mu_L^2}{\rho_l^2 \cdot g} \right)^{1/3}$	liquid phase Sherwood number

AUTHOR INDEX

American Institute of Chemical Engineers
3 Park Avenue
New York, New York 10016-5991